Introduction to PSpice®
Using OrCAD® Lite

ELECTRIC CIRCUITS
ELEVENTH EDITION

Introduction to PSpice®
Using OrCAD® Lite

E L E C T R I C   C I R C U I T S
ELEVENTH EDITION

JAMES W. NILSSON · SUSAN A. RIEDEL
Professor Emeritus, Iowa State University   *Marquette University*

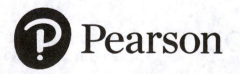 Pearson

330 Hudson Street, NY, NY 10013

Senior Vice President Courseware Portfolio Management, Engineering, Computer Science, Mathematics, Statistics, and Global Editions: Marcia J. Horton
Director, Portfolio Management, Engineering, Computer Science, and Global Editions: Julian Partridge
Specialist, Higher Ed Portfolio Management: Norrin Dias
Portfolio Management Assistant: Emily Egan
Managing Producer, ECS and Mathematics: Scott Disanno
Senior Content Producer: Erin Ault
Manager, Rights and Permissions: Ben Ferrini
Operations Specialist: Maura Zaldivar-Garcia
Inventory Manager: Bruce Boundy
Product Marketing Manager: Yvonne Vannatta
Field Marketing Manager: Demetrius Hall
Marketing Assistant: Jon Bryant
Project Manager: Rose Kernan
Cover Design: Black Horse Designs
Cover Art: © Leonardo Ulian, Matrix board series 06 - Resistance by abstraction, 2017.
Composition: Integra Publishing Services
Cover Printer: Courier/North Chelmsford
Printer/Binder: Courier/North Chelmsford

ISBN-10:   0-13-473933-7
ISBN-13: 978-0-13-473933-5

CONTENTS

Preface

ABOUT THIS MANUAL

Introduction to PSpice® Using OrCAD® Lite 17.2 uses OrCAD Lite PSpice A/D, Release 17.2 (herein after referred to as PSpice) as part of an introductory course in electric circuit analysis based on the textbook *Electric Circuits*. This supplement focuses on three things: (1) learning to draw and simulate linear circuits using PSpice, (2) using circuit models of basic devices such as op amps and (3) challenging simulation results to verify their accuracy. PSpice supports simulating networks with integrated circuit devices, so its range of application goes well beyond the topics covered here. Even though we do not exploit the full power of PSpice, we introduce you this widely used simulation program so you can verify your analysis of electric circuits using computer simulation.

To use PSpice you need to learn a new technical vocabulary and a number of specialized techniques. Hence, we designed this supplement to stand on its own as an instructional unit. Separating this material from the parent textbook also is a service to you: the portable format makes it easier to use this supplement at a computer terminal.

You can use PSpice to solve many of the textbook's Assessment Problems and Chapter Problems. Those Chapter Problems that are particularly suited to PSpice simulation are marked with a PSPICE icon in the textbook. You can use PSpice to check your analytical solutions to Chapter Problems, or to further explore the behavior of an interesting circuit.

Table 1: Relating Topics in this Supplement to Topics in the Text

MANUAL CHAPTER	TOPIC	TEXT CHAPTER
2–3	DC analysis	2–4
4	Thévenin Equivalents	4
4	Tolerance and Sensitivity	4
5	Op amps	5
6	Natural Response	7–8
7	Step response	7–8
7	Switches	7
8	Varying component values	8
9	AC Steady-state analysis	9
10	Linear Transformers	9–10
10	Ideal transformers	9–10
11	AC Power	10–11
12	Frequency response	13–15
12	Filter Design	14–15
13	Pulsed sources	16
13	Fourier series analysis	16

INTEGRATING PSPICE INTO INTRODUCTORY CIRCUITS COURSES

Although some circuits courses cover PSpice as an independent topic, many instructors prefer to integrate computer solutions with the course. To support integration, topics in this supplement appear in the same order in which they are presented in the text. Table 1 summarizes the relationship between the supplement and the textbook.

ABOUT PSPICE

SPICE is a software application that supports circuit design and simulation. SPICE is the acronym for a Simulation Program with Integrated Circuit Emphasis. The Electronics Research Laboratory of the University of California developed SPICE and made it available to the public in 1975.

Many different software packages implement SPICE on personal computers or workstations. Among them is OrCAD PSpice A/D, from Cadence. PSpice's popularity can be attributed to many factors, including its user-friendly interface, extensions to SPICE that support custom circuit component models, and its no-cost basic version. This manual focuses on how to use OrCAD Lite Release 17.2, and all examples were produced using this release. If you are using a different version of PSpice, or another package that implements SPICE, your use interface may differ from what you see in this supplement.

We limit the applications of PSpice to the types of circuit problems discussed in the textbook. Although PSpice is a general-purpose program designed for a wide range of circuit simulation — including the simulation of nonlinear circuits, transmission lines, noise and distortion, digital circuits, and mixed digital and analog circuits — here we discuss the use of PSpice only for dc analysis, transient analysis, and steady-state sinusoidal (ac) analysis. You can explore the extensive documentation on the Cadence website to find out more about the features of PSpice that are not discussed in this supplement.

To download OrCAD Lite, which is free and not time-limited, go to

http://www.orcad.com/resources/download-orcad-lite.

ASSESSMENT

In order to assess your understanding of the material in each chapter of this manual, we have included a set of suggested problems from the Eleventh Edition of *Electric Circuits* at the conclusion of each chapter. You are encouraged to solve these problems by hand, and then use PSpice to verify your solutions or to create plots that enable you to visualize the circuit behavior.

JAMES W. NILSSON
SUSAN A. RIEDEL

Chapter 1

A FIRST LOOK AT PSPICE

The general procedure for using PSpice consists of three basic steps:

1. Describe the circuit by drawing it in schematic form.
2. Specify the type of analysis and run the simulation.
3. Print or plot the simulation results.

In this first look, we develop a simple example that illustrates these steps.

1.1   DRAWING THE CIRCUIT

To begin, run the OrCAD Capture program, and select New Project from the Start Page, as shown in Fig. 1. This activates the New Project dialog box, shown in Fig. 2. Enter a meaningful project name and the desired directory for your project files and select the Analog Mixed-Mode Project Wizard, which will guide you through the rest of the process.

You can create your PSpice project based on an existing project, or create a blank project. Figure 3 shows the dialog for making this choice. When you exit this dialog, you are back in Capture, but now have a circuit layout grid and the Capture toolbar, as shown in Fig. 4.

Next, create your circuit schematic. Begin by selecting the menu option Place/Part, or by clicking on the second vertical toolbar button, as shown in Fig. 4. This activates the Place Part dialog. If you know the name of the library containing the part you want, highlighting the library name will display only the parts contained in that library. If the library

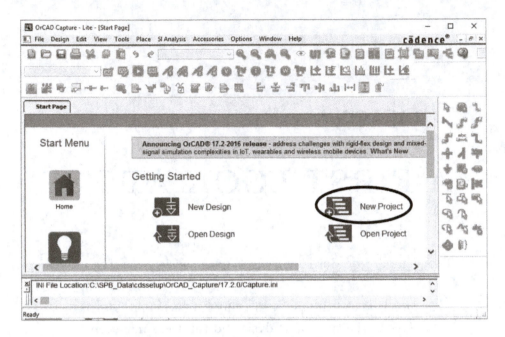

Figure 1: The Start Page for OrCAD Capture.

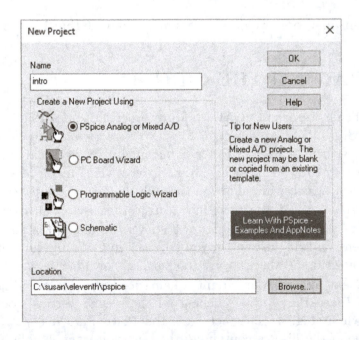

Figure 2: The New Project dialog box.

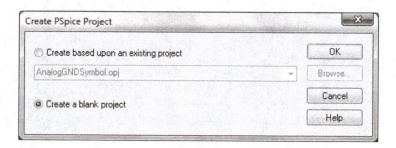

Figure 3: The dialog window for creating a PSpice project.

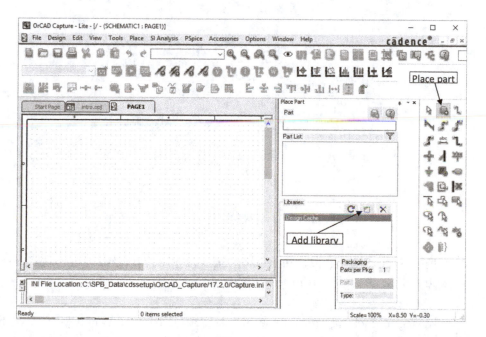

Figure 4: The circuit layout grid and the Place Part dialog.

is not shown, add it by clicking on the Add Library icon to display a list of library files, shown in Fig. 5. Usually the libraries you need are in the Cadence/SPB_17.2/tools/capture/library/pspice directory. As you can see in Fig. 5, we selected the part named VDC, a dc voltage source, from the library named Source. If you are not sure which library contains the part you need, you can highlight all of them to list all of the available parts. Clicking on a part name displays its schematic (see Fig. 5), so you can verify that this is the part you want. Double-click on the name of the part when you are ready to place the part in the schematic.

Next, place one or more copies of the part you have selected on the layout grid. You can change the orientation of the current part using Control-R or

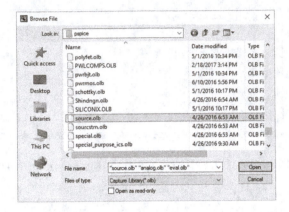

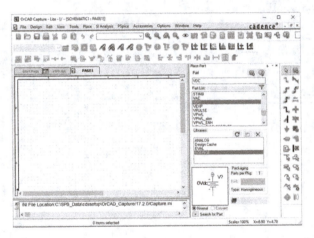

Figure 5: Selecting a part: adding the library containing the part (top); selecting the part from the library (bottom).

by right clicking the mouse and selecting rotate. When you have finished placing as many copies of the current part as you need, right click on the mouse and select End Mode, as shown in Fig. 6.

Now, specify the attributes of the part (or parts) you have placed. Usually the only attribute you must specify is the part's value, which in this case is the value of the voltage supplied by the dc voltage source. To change this value, double click on it to access the Display Properties dialog for this part, shown in Fig. 7. We set the dc voltage source's value to 12V, or 12 volts. The number you specify can be an integer (4, 12, −8) or a real number (2.5, 3.14159, −1.414). Integers and real numbers can be followed by either an integer exponent (7E−6, 2.136E3) or a symbolic scale factor (7U, 2.136k). Table 2 summarizes the symbolic scale factors used in PSpice and their corresponding exponential forms. Letters immediately following

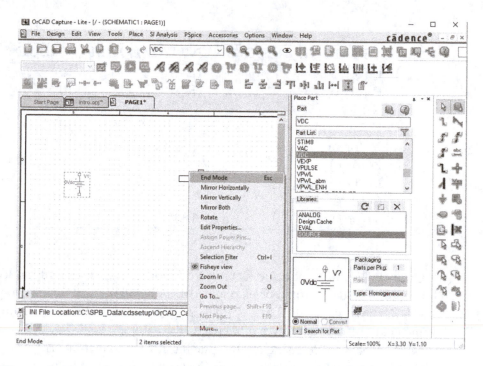

Figure 6: Place a dc voltage source on the layout grid, then right click and select End Mode to finish with this part.

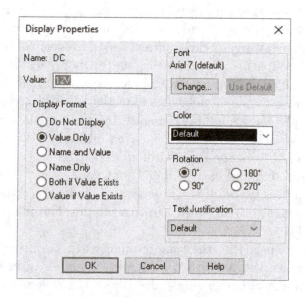

Figure 7: Use the Display Properties dialog window to set the dc voltage source value.

Table 2: PSpice Scale Factors

SYMBOL	EXPONENTIAL FORM	VALUE
F (or f)	1E−15	10^{-15}
P (or p)	1E−12	10^{-12}
N (or n)	1E−9	10^{-9}
U (or u)	1E−6	10^{-6}
M (or m)	1E−3	10^{-3}
K (or k)	1E3	10^{3}
MEG (or meg)	1E6	10^{6}
G (or g)	1E9	10^{9}
T (or t)	1E12	10^{12}

a number that are not scale factors are ignored, as are letters immediately following a scale factor. For example, 10, 10v, 10HZ, and 10A all represent the same number, as do 2.5m, 2.5MA, 2.5msec, and 2.5MOhms. Component values cannot contain any white space characters.

The process of locating a part, placing it on the layout grid, and setting its value is repeated for each part in your circuit. Figure 8 shows the circuit we want to simulate, which has a 12 V dc source and two resistors with values of 3 kΩ and 6 kΩ. Note that resistors are in the Analog library.

After selecting your components, placing them on the grid and specifying their values, you are ready to wire the circuit. You can select the menu option Place/Wire, or click on the Place wire vertical toolbar icon, as shown in Fig. 8. Wires are placed by clicking the mouse at the starting node for the wire, and clicking the mouse again at the ending node for the wire. Figure 9 shows the example circuit with all of the wires placed.

Complete the circuit schematic by adding the circuit ground. To do this, click on the vertical toolbar icon labeled GND (see Fig. 9) and select 0/CAPSYM from the Place Ground dialog, shown in Fig. 10. This is a very important step and easy to forget, but your circuit cannot be simulated by PSpice unless it contains a ground at node 0! Place the ground component on the layout grid, and wire it to your circuit. The completed circuit is shown in Fig. 11.

Before specifying the type of analysis to perform, let's briefly examine what PSpice will do with the circuit schematic. In order for PSpice to understand the circuit we have described, the schematic must be translated into a collection of statements that identify the circuit components, their attributes, and their topological connections. This collection of statements is called a netlist, and it is written to the output file before analysis begins.

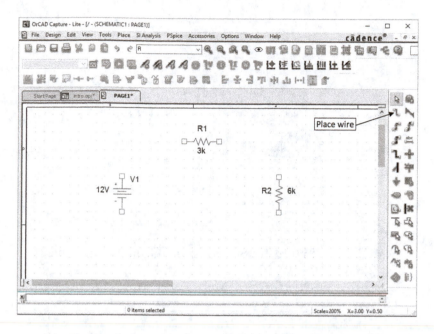

Figure 8: All circuit components are selected and placed on the layout grid.

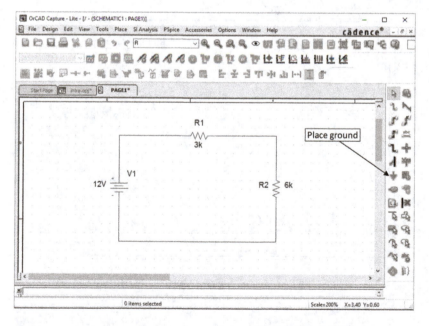

Figure 9: The circuit with all components wired together.

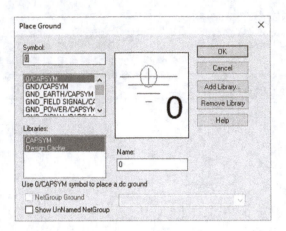

Figure 10: Selecting the Ground component.

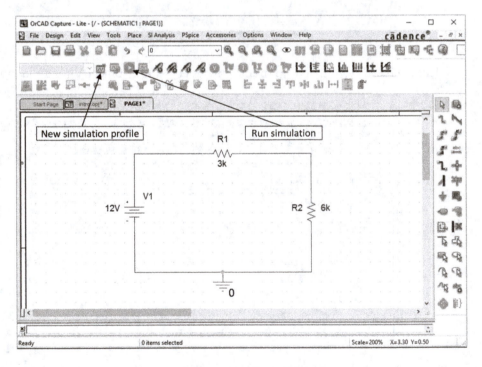

Figure 11: A completed circuit schematic.

Sometimes it is useful to look at the netlist, especially if there is an error in the schematic. Therefore, the Appendix includes a subset of the netlist syntax for your reference.

1.2 SPECIFYING THE TYPE OF CIRCUIT ANALYSIS

PSpice can perform several different types of analysis on a circuit. To select a name for the simulation, choose the menu option PSpice/New Simulation Profile, or click the icon indicated in Fig. 11, which will generate the dialog shown in Fig. 12. We've named this profile "bias", as we will specify a simple dc analysis that determines all of the node voltage values, a process known as "biasing". Click the Create button and the Simulation Settings dialog window will appear, as shown in Fig. 13. Select the type of analysis using the Analysis Type dropdown list on the Analysis tab. We have selected Bias Point, as you can see from Fig. 13. You can also see that there are many other tabs in the Simulation Settings dialog that can be used to further configure the analysis. Usually the default values suffice, as they do in this example. Once you select the analysis type, click on OK to return to your schematic.

Circuit analysis begins when you select the menu option PSpice/Run or by clicking the Run icon shown in Fig. 11. Once the analysis concludes you can see the bias point analysis results by clicking on the V, I, and W icons highlighted in Fig. 14 to display the node voltages, component currents and component power values.

When the simulation concludes a second window appears, as shown in Fig. 15. The window is divided into five sections. The four sections on the bottom, from left to right, are a command window, a parameter value monitoring window, a simulation progress window, and a results value window. The section on the top is used to print or plot the simulation results and is discussed in more detail next.

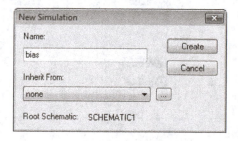

Figure 12: Creating a new simulation profile.

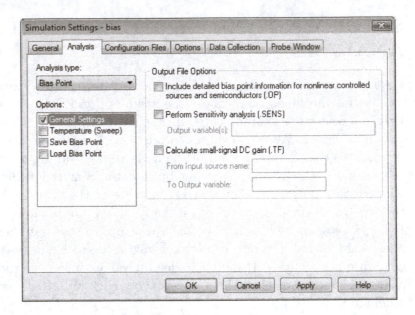

Figure 13: The Simulation Settings dialog window, used to specify the type of circuit analysis.

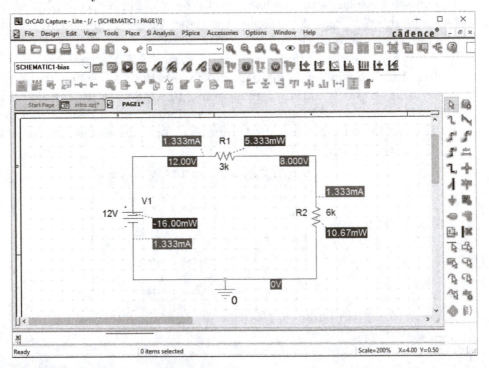

Figure 14: After bias analysis completes select the V, I, and W icons, highlighted in the horizontal toolbar, to display the voltage, current, and power.

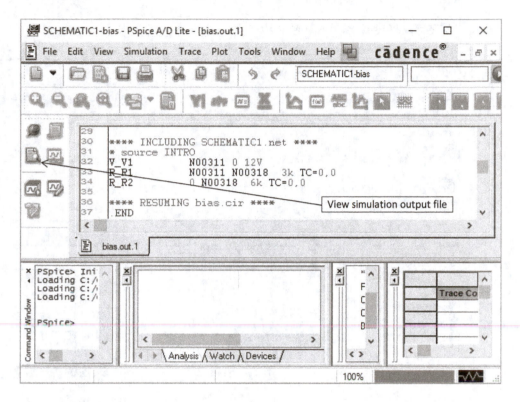

Figure 15: Bias analysis output window, displaying the output file produced during simulation.

1.3 SIMULATION RESULTS

Generally, the simulation results can be a text file, a plot, or both. In this simple example we cannot examine a plot, as no parameters have been varied. To view the output file click on the vertical toolbar icon indicated in Fig. 15. Near the top of this output file is the netlist describing the circuit schematic in a form that PSpice can understand. The netlist consists of the three parts we drew: a 12 V dc voltage source between nodes N00311 and 0 (the required ground node); a 3 kΩ resistor between nodes N00311 and N00318; and a 6 kΩ resistor between nodes N00318 and 0.

Scrolling further down in the output file, we see the results of the bias analysis, including the value of the voltage at each of the non-ground nodes, as shown in Fig. 16. The voltage at node N00311 is 12 V, which we expected since the 12 V dc voltage source is between node N00311 and ground.

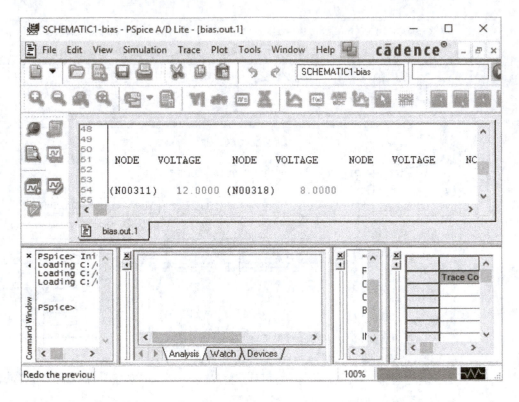

Figure 16: Scroll through the simulation output file to display the node voltages.

The voltage at node N00318 is 8 V, which we also should have expected, since the circuit we created is a simple voltage divider and

$$v_{6k\Omega} = \frac{6}{3+6}(12) = 8 \text{ V}.$$

This completes the simple example. The remainder of this supplement describes the various types of components used in linear circuits, the other types of analysis available in PSpice, and the use of Probe to plot the results of circuit simulation. Note that historically, the plotting functions in PSpice were collectively known as "Probe". We will continue to call these functions Probe in this text.

ASSESSMENT

◆ Use PSpice to solve the following problems: 3.19(a), 3.28, 3.30, and 3.32(a).

Chapter 2

SIMPLE DC CIRCUITS

In this chapter we look at circuits containing independent sources, dependent sources, and resistors. We continue to analyze such circuits using a simple bias calculation.

2.1 INDEPENDENT DC SOURCES

Parts that model independent dc voltage and current sources are found in the Source library, and are named VDC and IDC, respectively. After placing the independent sources in the layout grid, double click on the value of the source to specify the desired value, which can be either a positive or a negative number. We saw an example of VDC in Chapter 1.

2.2 DEPENDENT DC SOURCES

Placing and wiring dependent sources can be a little tricky, as they each require four wired connections—two that connect the source into the circuit, and two that connect the controlling voltage or current into the source.

We start with a voltage-controlled voltage source. A circuit containing this type of dependent source, together with an independent dc current source and several resistors, is shown in Fig. 17. The part name for the voltage-controlled voltage source is E, and is found in the Analog library. The Place Part dialog for this part is shown in Fig. 18. The schematic representation of this part, as seen in the figure, has four connection points. The two points on the right, labeled + and − inside a circle, are used to wire this component into the circuit between the 6 A current source and the 4 Ω resistor (see Fig. 17). The other two nodes are attached to the circuit

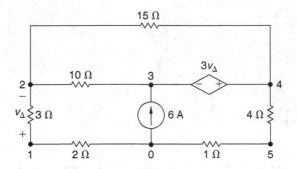

Figure 17: Circuit used to illustrate a voltage-controlled voltage source.

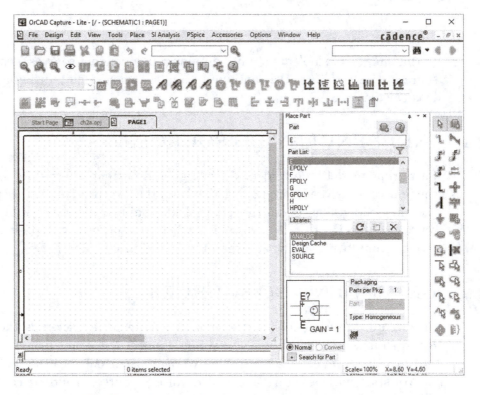

Figure 18: The Place Part dialog for a voltage-controlled voltage source.

and measure the controlling voltage. These two nodes must be connected in parallel with the component whose voltage is the controlling voltage. This is the 3 Ω resistor in Fig. 17, where the controlling voltage v_Δ is defined.

The circuit schematic with all of the parts placed but no connecting wires, is shown in Fig. 19. Note that we have rotated the voltage-controlled voltage source to prepare it for wiring. Before we wire the schematic, we

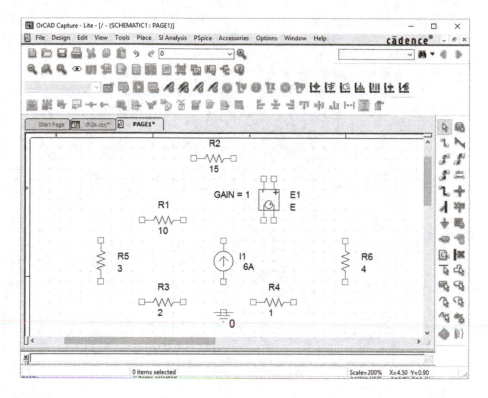

Figure 19: Parts placed for the schematic describing the circuit in Fig. 17.

specify the gain of the voltage-controlled voltage source. To do this, click on the voltage-controlled voltage source (not its labels), which will highlight this part and place a dashed box around it. Then select the menu option Edit/Properties to invoke the Property Editor. Scroll horizontally in the property spreadsheet to find the column labeled GAIN, click on the value of the gain and change it to 3, as shown in Fig. 20.

Now the parts can be wired and the ground added to complete the schematic, which is shown in Fig. 21. Make sure your wired circuit maintains the voltage polarities shown in the circuit diagram. You might need to zoom in on the schematic as you wire in order to see the polarity markings on the dependent source. Use the zoom toolbar button or the menu option View/Zoom.

The schematic representations for the remaining dependent dc sources are shown in Fig. 22 (on p. 18). These dependent dc sources are also in the Analog library. The current-controlled current source has the part name F, the voltage-controlled current source has the part name G, and the current-controlled voltage source has the part name H. Wire the dependent current

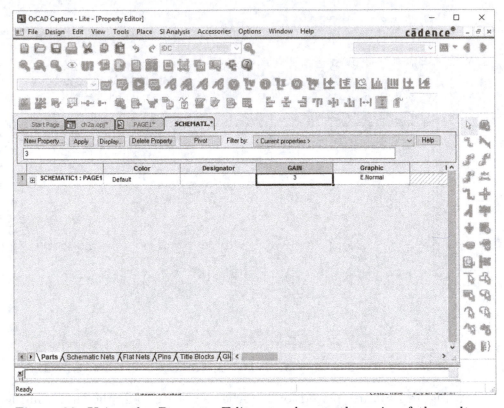

Figure 20: Using the Property Editor to change the gain of the voltage-controlled voltage source.

sources into a circuit using the terminals attached to the circle with the arrow, shown on the right of the schematic for parts F and G. The dependent sources that are controlled by currents (parts F and H) have a large arrow on the left side of the part (see Fig. 22). These terminals must be wired in series with the component through which the controlling current is defined. We will see an example of a circuit with a current-controlled dependent source in the next section.

2.3   RESISTORS

Chapter 1 performed a dc bias analysis on a circuit containing resistors and a dc voltage source. There, we saw that the model of a resistor has the part name R, and this part is found in the Analog library. Once we place the resistors in our schematic, double clicking on their resistance value allows

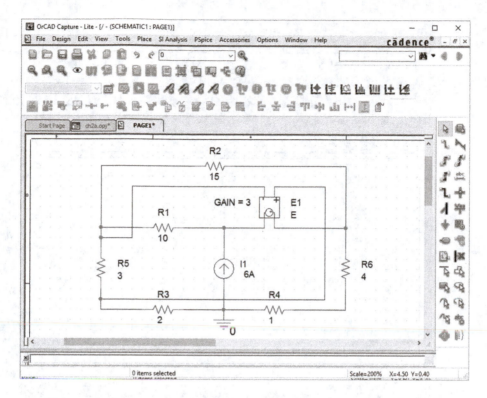

Figure 21: The completed schematic for the circuit in Fig. 17.

us to modify the resistance. The following example performs the bias point analysis of a circuit containing an independent source, a dependent source, and some resistors.

EXAMPLE 1

a) Use PSpice to find the voltages v_a and v_b for the circuit shown in Fig. 23. Enter the gain of the dependent source as 20.384615 (=265/13) in the Property Editor.

b) Use PSpice to find (1) the total power dissipated in the circuit, (2) the power supplied by the independent current source, and (3) the power supplied by the current-controlled voltage source.

SOLUTION

a) The circuit shown in Fig. 23 is represented as a PSpice schematic in Fig. 24. Note the wiring of the current-controlled voltage source. After bias point analysis, display the node voltages and the component power

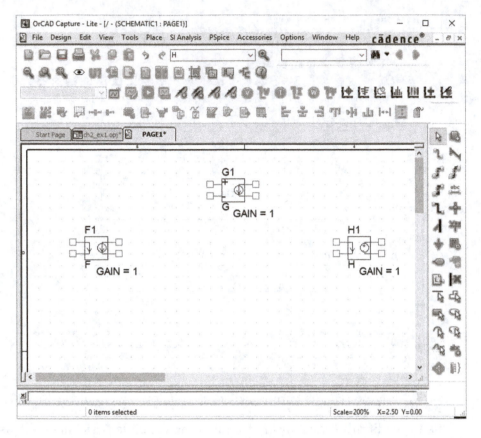

Figure 22: The schematic components representing a current-controlled current source (part name F), a voltage-controlled current source (part name G), and a current-controlled voltage source (part name H).

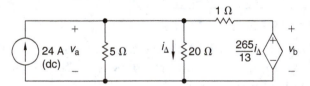

Figure 23: Circuit for Example 1.

values using the V and W icon in the horizontal toolbar, as shown in Fig. 25. We see that

$$v_a = 104.00 \text{ V};$$
$$v_b = 106.00 \text{ V}.$$

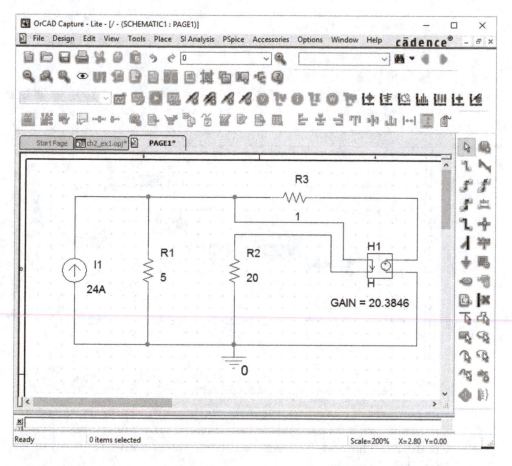

Figure 24: The schematic representation of the circuit in Fig. 23.

b) The total power dissipated is the sum of the positive power values in Fig. 25. The power supplied by the two sources is also shown in Fig. 25. Therefore,

1. $\sum P_{\text{dis}} = 2163.2 + 540.8 + 4 = 2708$ W.

2. $P_{24\text{A}}(\text{supplied}) = 2496$ W.

3. $P_{\text{H1}}(\text{supplied}) = 212$ W.

Note that the total power dissipated equals the total power supplied.

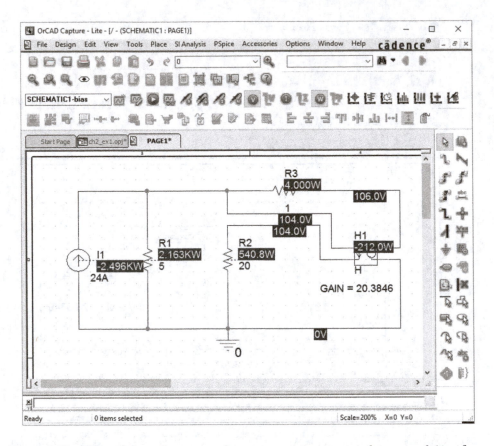

Figure 25: The node voltages and component power values resulting from Bias Point analysis.

ASSESSMENT

♦ Use PSpice to solve the following problems: 4.6, 4.11, 4.13, 4.21, and 4.28.

Chapter 3

DC SWEEP ANALYSIS

In the first two chapters, we used PSpice to compute the bias point (the dc operating point) for several circuits. PSpice always calculates the operating point because it is needed before proceeding with other types of analysis.

In this chapter, we explore DC Sweep analysis, which enables us to simulate a circuit for a range of dc current or voltage source values.

3.1 SWEEPING A SINGLE SOURCE

Example 2 uses DC Sweep analysis to calculate an output voltage and current for different source voltage values.

EXAMPLE 2

For the circuit shown in Fig. 26, use PSpice to find the values of i_o and v_o when v_g varies from 0 to 100 V in 10 V steps.

SOLUTION

The circuit shown in Fig. 26 is specified in schematic form in Fig. 27. Several new parts appear in this schematic. The voltage source is modeled by the part VSRC from the Source library, and the current source is modeled by the part ISRC from the Source library. Note that unneeded labels for these parts are not shown in the figure. We could also have used the parts VDC and IDC for this example. To specify the current source value, click the part to highlight it and select the menu option Edit/Properties. Scroll over to the column labeled DC and type in the

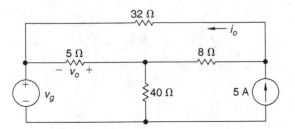

Figure 26: The circuit for Example 2.

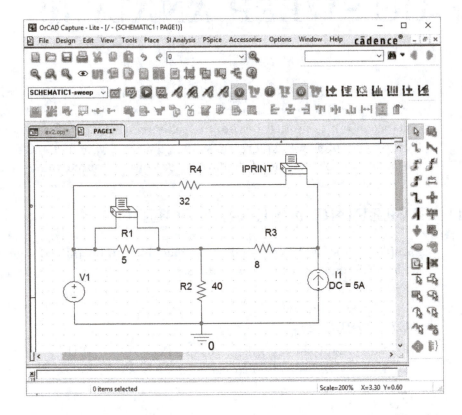

Figure 27: The schematic representation of the circuit in Fig. 26.

value of the current source in amps, 5. There is no need to specify the value of the voltage source, as you will see.

The schematic includes printers, which write the output current and output voltage values to the output file. The current printer, Iprint, is located in the Special library. Connect the current printer in series with the component whose current you want to measure, so it acts like an ammeter. It is important to orient the current printer so the current exits the side of the symbol with the − sign (see Fig. 27). To activate the printer, highlight it and select the menu option Edit/Properties. Scroll over to the column labeled DC and enter the value Y (for "yes"), as shown in Fig. 28.

There are two voltage printer parts, both in the Special library. Vprint1 prints the voltage at a single node (with respect to ground), and Vprint2 prints the voltage between two nodes. We have used Vprint2 in the schematic shown in Fig. 27. This part must be wired into the circuit in parallel with the component whose voltage is being measured, just like a voltmeter. Again, the orientation of this part is important, so place the symbol such that the node labeled − in the circuit schematic

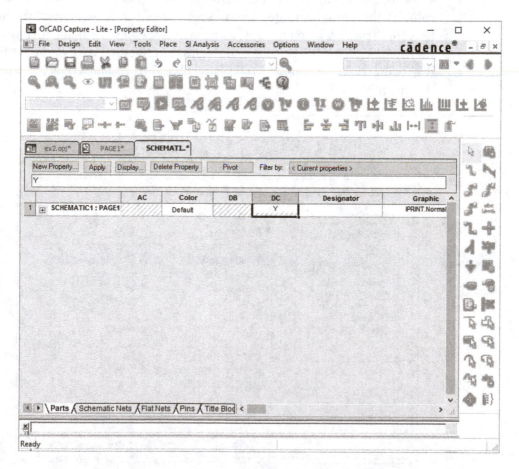

Figure 28: The Property Editor for the current printer Iprint.

is wired to the side of the printer symbol with the $-$ sign (see Fig. 27). You must edit the properties of the voltage printer and enter Y as the DC value to activate the printer.

Once the schematic is complete, create a new simulation profile for this circuit and select DC Sweep as the analysis type. Enter the name of the part whose values will be swept, in this case the voltage source V1, and the Start Value, Final Value, and Increment, as shown in the Simulation Settings dialog in Fig. 29. Run the PSpice simulation and examine the output file, a portion of which is shown in Fig. 30, where you can see the effect of changing the source voltage on the output voltage.

We can also depict the results of the DC Sweep analysis graphically. In fact, after PSpice has completed the simulation, you will notice that a graph appears in the upper section of the simulation window, with the value of the source voltage already selected as the independent variable. Use the Trace/Add menu option to specify which variable to plot. As shown in Fig. 31, we selected the voltage difference across the resistor R1, using the expression V1(R1:2) $-$ V1(R1:1). The resulting plot is shown in Fig. 32.

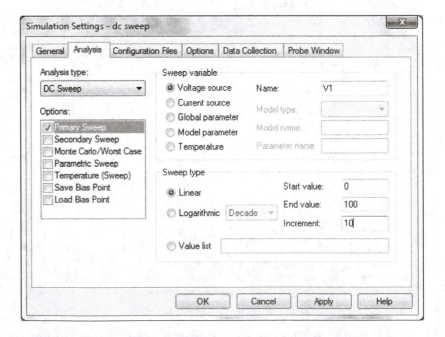

Figure 29: The Simulation Settings dialog for the DC Sweep analysis of the schematic in Fig. 27.

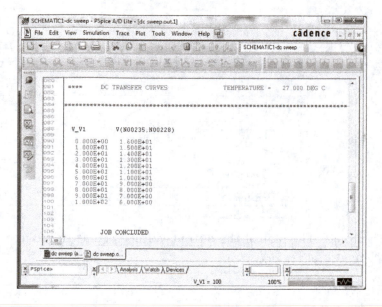

Figure 30: The results of DC Sweep analysis of the schematic in Fig. 27.

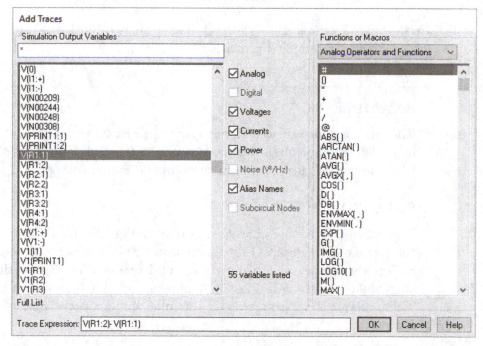

Figure 31: Selecting a variable to plot in Probe.

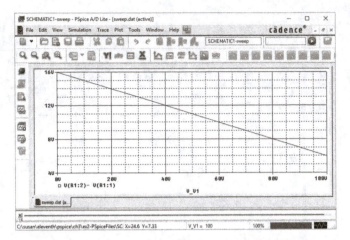

Figure 32: The plot of the output voltage versus the source voltage for the schematic in Fig. 27.

3.2   SWEEPING MULTIPLE SOURCES

Now we alter Example 2 to sweep through values of both the current and voltage sources. Example 3 presents the results.

EXAMPLE 3

The current source in the circuit shown in Fig. 26 varies from 0 to 5 A in 1 A steps. The voltage source varies from 0 to 100 V in steps of 20 V. Plot the output current as a function of the source voltage.

SOLUTION

The schematic is the same as the one in Fig. 27. In Capture, select the menu option PSpice/Edit Simulation Profile to specify a Secondary Sweep, the additional sweep variable and its values. The Simulation Settings dialog is shown in Fig. 33. The resulting plot is shown in Fig. 34. There are six separate traces on this plot, one for each value of source current.

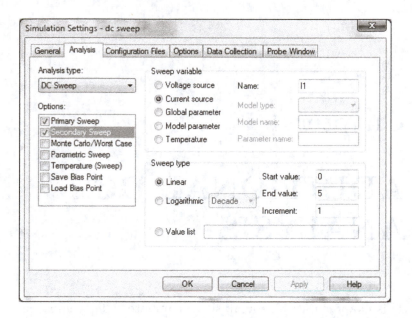

Figure 33: Adding a secondary sweep to the DC Sweep analysis of the schematic in Fig. 27.

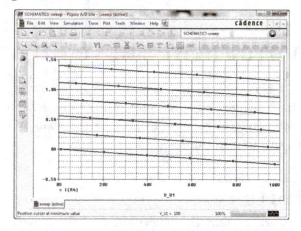

Figure 34: Plot of output current as source voltage and source current are varied, for the schematic in Fig. 27.

ASSESSMENT

♦ Use PSpice to solve the following problems: 4.54 and 4.55.

Chapter 4

ADDITIONAL DC ANALYSIS

This chapter uses bias analysis to compute a Thévenin equivalent, and dc analysis to examine the sensitivity of circuit output variables to changes in circuit parameter values and to calculate the effect of resistor tolerances on circuit outputs.

4.1 COMPUTING THE THÉVENIN EQUIVALENT

You can use bias analysis to calculate the open-circuit voltage and the short-circuit current at a designated pair of nodes in a circuit, and thus compute the Thévenin equivalent with respect to that pair of terminals. Example 4 illustrates this application.

EXAMPLE 4

Use PSpice to find the Thévenin equivalent with respect to terminals a and b for the circuit shown in Fig. 35.

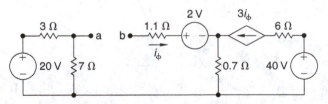

Figure 35: The circuit for Example 4.

SOLUTION

A schematic created from the circuit in Fig. 35 is shown in Fig. 36. We have selected node b in Fig. 35 as the reference node. To compute the short-circuit current, we inserted a resistor between nodes a and b (R2) whose value is much smaller than the value of other resistors in the circuit, in this case 10^{-6} Ω, effectively creating a short circuit. After Bias analysis, display the component currents on the circuit schematic, using the I icon in the horizontal toolbar. The results, shown in Fig. 37, confirm that the short-circuit current, which is the current through the R2 resistor, is 2 A.

To compute the open-circuit voltage, connect a resistor that is much larger than the other resistors in the circuit—say, 10^6 Ω for this circuit—between nodes a and b, effectively creating an open circuit. The resulting schematic is shown in Fig. 38. After Bias analyisis, display the node voltages on the circuit schematic using the V icon in the horizontal toolbar, as shown in Fig. 39.

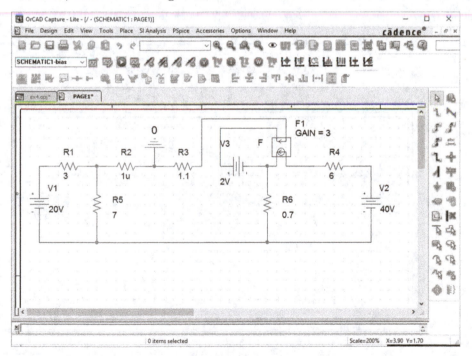

Figure 36: The schematic for the circuit in Fig. 35, configured to calculate the short-circuit current.

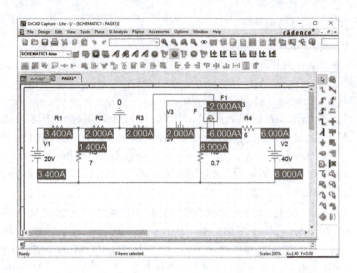

Figure 37: The bias point currents, showing that the current in the short circuit is 2 A.

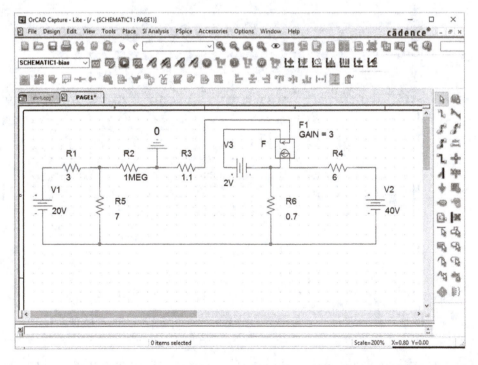

Figure 38: The schematic for the circuit in Fig. 35, configured to calculate the open-circuit voltage.

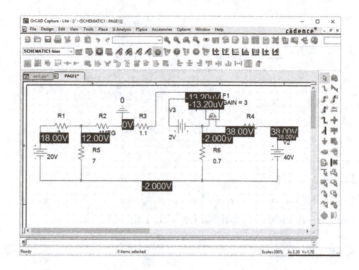

Figure 39: The bias point voltages, showing that the voltage across the open circuit is 12 V.

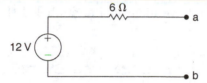

Figure 40: The Thévenin equivalent for Example 4.

The open-circuit voltage, which is the voltage across the R2 resistor, is 12 V. Therefore, the Thévenin resistance is 12V/2 A = 6 Ω and the Thévenin equivalent circuit is as shown in Fig. 40.

4.2 SENSITIVITY ANALYSIS

Sensitivity analysis finds the dc small-signal sensitivities of each specified output variable with respect to *every* circuit parameter. Circuits with a large number of elements generate tremendous amounts of output data when the sensitivity of more than one output variable is of interest. Sensitivity analysis can be included in the bias point analysis, and requires you to identify the voltages or currents in the circuit whose sensitivities are to be calculated. Example 5 predicts the behavior of an unloaded voltage-divider circuit using sensitivity analysis.

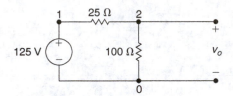

Figure 41: The circuit for Example 5.

EXAMPLE 5

Use PSpice to study the sensitivity of the output voltage v_o in the voltage-divider circuit shown in Fig. 41.

SOLUTION

Because we have three elements in the circuit, PSpice calculates the sensitivity of v_o with respect to the independent voltage source, the 25 Ω resistor, and the 100 Ω resistor. The PSpice schematic is shown in Fig. 42. Note that we have identified and labeled the output node of interest for

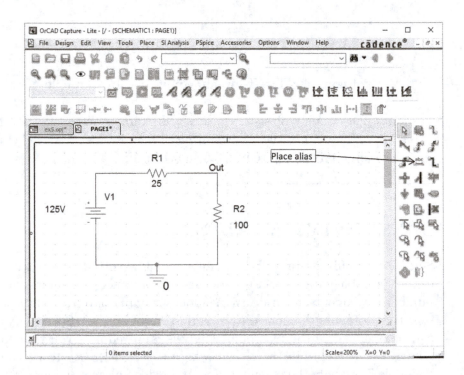

Figure 42: The schematic representation of the circuit in Fig. 41.

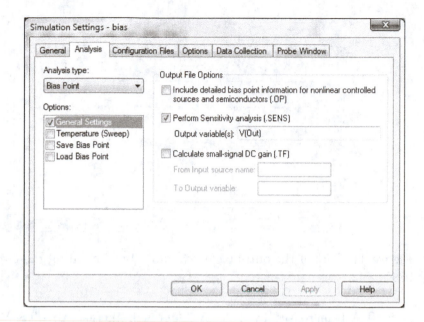

Figure 43: Specifying sensitivity analysis in PSpice.

sensitivity analysis. Do this by placing an alias at the node above the R2 resistor using the vertical toolbar button shown in Fig. 42.

To perform sensitivity analysis, select Bias Point as the analysis type, and check the sensitivity analysis box, as shown in Fig. 43. We have selected V(Out) as the variable of interest. PSpice prints the data associated with sensitivity analysis in the output file. The sensitivity analysis output is shown in Fig. 44. From the simple dc analysis of the circuit, $v_o = V(\text{Out}) = 100$ V when R1, R2, and V1 are at their nominal values. From the sensitivity data, we deduce that

1. If R1 increases by 1 Ω, V(Out) will decrease by 0.8 V—that is, $V(\text{Out}) = v_o = 99.2$ V;

2. If R1 increases by 1%, v_o will decrease by 0.2 V to 99.8 V;

3. If R2 increases by 1 Ω, V(Out) will increase by 0.2 V—that is, $V(\text{Out}) = v_o = 100.2$ V;

4. If R2 increases by 1%, v_o will increase by 0.2 V to 100.2 V;

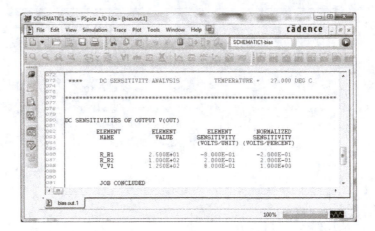

Figure 44: Part of the output file generated by simulating the circuit shown in Fig. 42.

5. If V1 increases by 1 V, V(Out) will increase by 0.8 V—that is, V(Out) $= v_o = 100.8$ V; and

6. If V1 increases by 1%, v_o will increase by 1 V to 101 V.

Because we have a linear circuit, the principle of superposition applies, and therefore we may superimpose simultaneous effects. For example, let's assume that R1 increases by 1 Ω, R2 decreases by 1%, and V1 increases by 0.5 V. The effect on v_o, or V(Out), will be $v_o = 100 - 0.8 - 0.2 + 0.4 = 99.4$ V.

4.3   SIMULATING RESISTOR TOLERANCES

We examined the effect of choosing a particular component value on the output of the circuit using sensitivity analysis. But we might also be interested in simulating the effect of variation in component values, called tolerances, on the output of the circuit. The PSpice model for resistors assumes that the resistor is perfect, never varying from its desired resistance value. In the next example, we alter the model of the resistor to specify a realistic tolerance value. With this more realistic component in our schematic, we then perform worst-case analysis to discover how resistor tolerance impacts the output of our circuit.

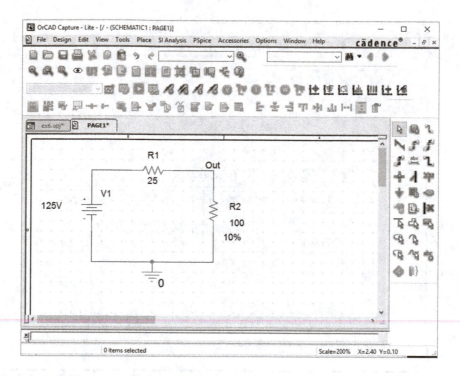

Figure 45: The schematic representation of the circuit in Fig. 41, using a resistor model with 10% tolerance.

EXAMPLE 6

Replace the 100 Ω resistor in the circuit in Fig. 41 with a resistor having the same value and a 10% tolerance. Use PSpice to discover the range of output voltage values that you can expect with this more realistic model of a resistor.

SOLUTION

The schematic for the circuit in Fig. 41 is shown in Fig. 45. Note that the R2 resistor includes a 10% tolerance value. We used the Property Editor for the R2 resistor to set its tolerance. Find the column labeled Tolerance within the Property Editor and enter 10% as its value, as shown in Fig. 46. Then click the Display button to display the tolerance value.

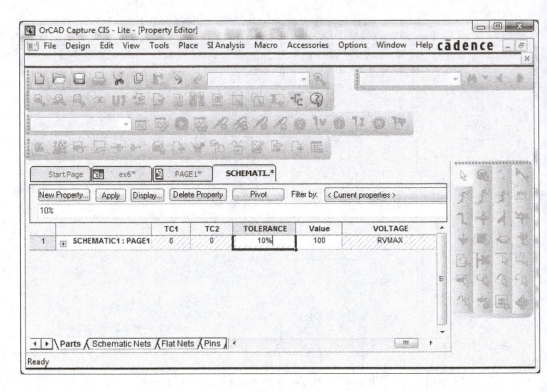

Figure 46: The Property Editor with a 10% tolerance for the R2 resistor in Fig. 45.

Next we create a new Simulation Profile, choosing DC Sweep analysis. Set both the minimum and maximum values to sweep the input voltage, V1, to 125 V, with an increment of 1 V. This screen is not shown. Then choose the Monte Carlo/Worst Case option, seen in Fig. 47, select the Worst Case/Sensitivity button, identify the output variable, and choose to vary devices that have device tolerances only. Once these settings have been specified, click on the More Settings button, activating the dialog shown in Fig. 48. We perform two worst-case analyses—one to determine the maximum worst-case output voltage, and the other to determine the minimum worst-case output voltage. The parameters for the maximum worst-case analysis are shown in Fig. 48, and the relevant portion of the output file is shown in Fig. 49. This output shows that if the resistor R2 is 10% larger than its nominal value of 100 Ω, the output voltage will be 101.85 V. You can confirm this result using the voltage division equation for the circuit in Fig. 41 when the R2 resistor is 110 Ω.

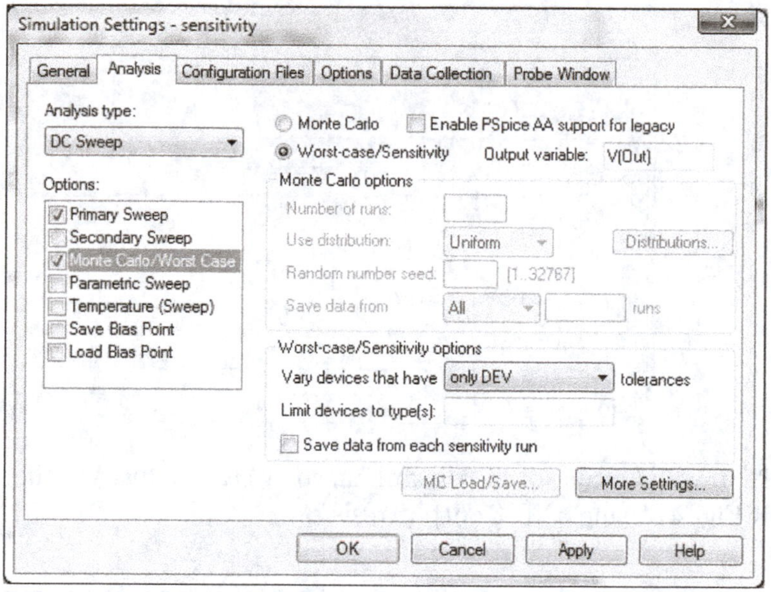

Figure 47: The first step in identifying parameters for worst-case analysis of resistor tolerance.

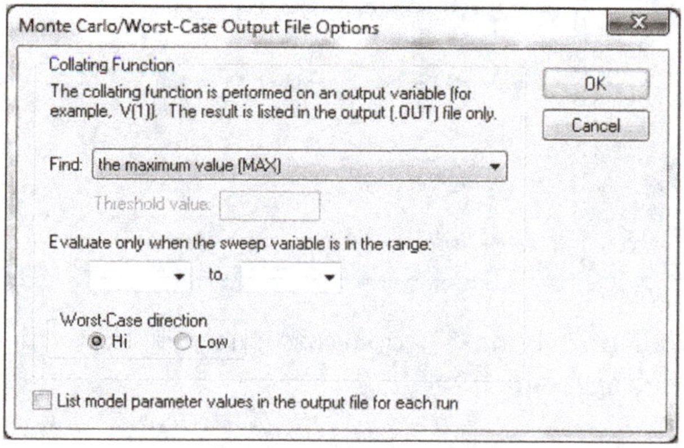

Figure 48: The second step in identifying parameters for worst-case analysis of resistor tolerance.

The second worst-case analysis is specified using the parameters in Fig. 50 and the results appear in Fig. 51. Now we see that if the resistor R2 is 10% smaller than its nominal value of 100 Ω, the output voltage will be 97.826 V. You can confirm this result using the voltage division equation for the circuit in Fig. 41 when the R2 resistor is 90 Ω. Thus,

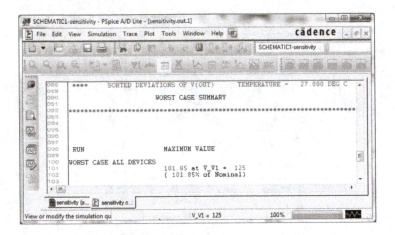

Figure 49: Output of the maximum worst-case analysis of the voltage divider in Fig. 41 using a 10% output resistor.

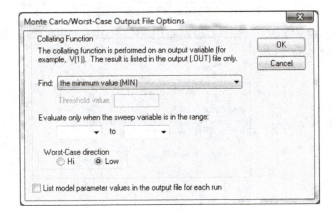

Figure 50: Specifying a worst-case analysis that calculates the minimum output voltage value.

when the R2 resistor in Fig. 41 has a tolerance of 10%, the actual output voltage may range from 97.826 V to 101.85 V. If you were designing this circuit for an actual application, you would need to evaluate whether this variation around the designed value of 100 V is acceptable.

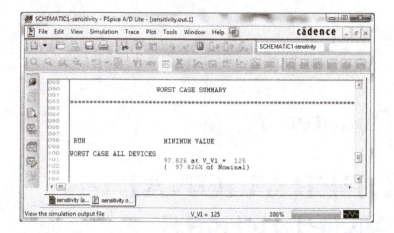

Figure 51: Output of the minimum worst-case analysis of the voltage divider in Fig. 41 using a 10% output resistor.

ASSESSMENT

♦ Use PSpice to solve the following problems: 4.64, 4.66, 4.67, 4.75, 4.105, 4.106, and 4.107.

Chapter 5

OPERATIONAL AMPLIFIERS

PSpice offers three options for describing an operational amplifier (op amp) within a circuit schematic.

- Model the op amp with resistors and dependent sources.

- Use the ideal op amp model in the PSpice Analog library.

- Use the model of an actual op amp from a PSpice library.

The library models are considerably more sophisticated than the simple dependent-source-based models and are designed to mimic the characteristics of actual op amps. In ORCAD Capture you can modify an existing op amp library model to exhibit the characteristics you desire. This modified model can then be used like any other PSpice circuit element, such as a resistor, and included in multiple locations within the circuit schematic. ORCAD Lite, used in this manual, does not support model editing, so no examples of this technique are presented.

The examples that follow illustrate each of the three options for incorporating an op amp in a circuit schematic.

5.1 MODELING OP AMPS WITH RESISTORS AND DEPENDENT SOURCES

A simple equivalent circuit for an operational amplifier is given in Fig. 52. Example 7 uses PSpice to analyze a circuit containing an op amp modeled with the simplified equivalent circuit in Fig. 52.

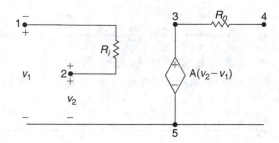

Figure 52: An equivalent circuit for an op amp.

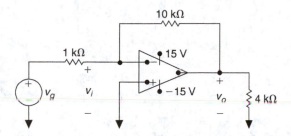

Figure 53: The circuit for Example 7.

EXAMPLE 7

The parameters for the op amp model used in the circuit of Fig. 53 are $R_i = 200$ kΩ, $A = 10^4$, and $R_o = 5$ kΩ.

a) Use PSpice to find v_i and v_o when $v_g = 1$ V(dc).

b) Compare the PSpice solution with the analytic solutions for v_i and v_o.

SOLUTION

a) The schematic for the circuit in Fig. 53 is given in Fig. 54. We used simple bias point analysis and then used the V icon to display the node voltages, as shown in Fig. 55. Therefore,

$$
\begin{aligned}
v_i &= V(\text{IN}) = 2.742 \text{ mV}; \\
v_o &= V(\text{OUT}) = -9.97 \text{ V}.
\end{aligned}
$$

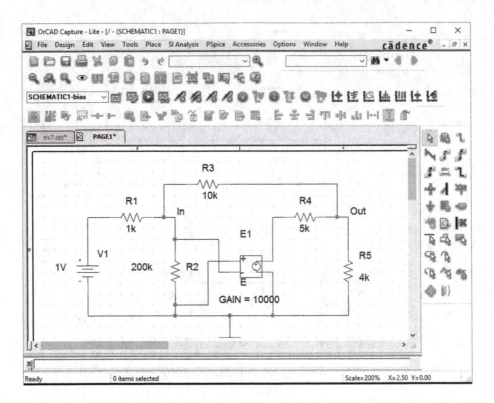

Figure 54: The schematic representation of the circuit in Fig. 53. using the equivalent circuit for the op amp from Fig. 52.

b) To solve for v_i and v_o, substitute the circuit in Fig. 52 for the op amp in Fig. 53. Then write two KCL equations, one at the node labeled v_i and the other at the node labeled v_o:

$$\frac{v_i - 1}{1000} + \frac{v_i}{200{,}000} + \frac{v_i - v_o}{10{,}000} = 0;$$

$$\frac{v_o - v_i}{10{,}000} + \frac{v_o + 10^4 v_i}{5000} + \frac{v_o}{4000} = 0.$$

Solving, we get $v_i = 2.742$ mV   and $v_o = -9.970$ V.

If, instead, we want to use an ideal op amp in the circuit of Fig. 53, we can alter the op amp model, making R_i and A very large and $R_o = 0$. To

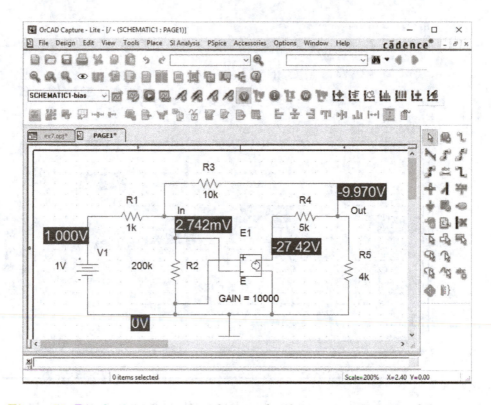

Figure 55: Displaying the node voltages after bias point analysis of the circuit in Fig. 54.

illustrate, we modify the schematic in Fig. 54, with R2 = 200 MΩ, A = 10^8, and R4 removed from the schematic entirely. The resulting node voltages after bias point analysis are shown in Fig. 56. The values of v_i and v_o from this simulation are

$$v_i = V(\text{IN}) = 100\text{nV} \approx 0 \text{ V};$$
$$v_o = V(\text{OUT}) = -10.0 \text{ V}.$$

To confirm these results, analyze the circuit shown in Fig. 53 assuming an ideal op amp.

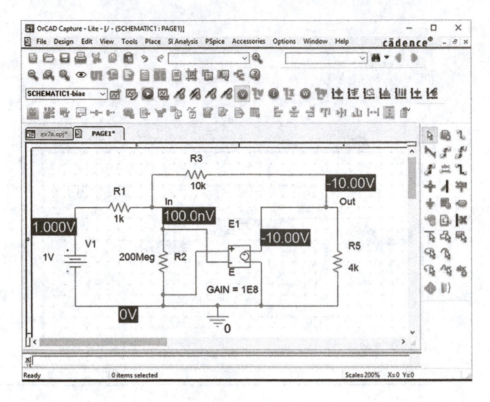

Figure 56: Displaying the node voltages after bias point analysis of the circuit in Fig. 54, using an ideal op amp model.

5.2   USING OP AMP LIBRARY MODELS

One of the most powerful features of PSpice is its library of models for many common electronic devices. The full version of PSpice has models for more than 34,000 devices. The evaluation version of PSpice (used to produce the examples in this supplement) has models for about 100 of the most commonly used devices. Among these models are several for op amps, including the μA741. They are all found in the Eval library. There is also a model for an ideal op amp, which is found in the Analog library. The next two examples repeat Example 7, but incorporate different op amp models in the circuit schematic for Fig. 53.

EXAMPLE 8

Repeat the analysis of the circuit shown in Fig. 53, using the ideal op amp model from the Analog library.

SOLUTION

Figure 57 shows a schematic for the circuit in Fig. 53 and uses the ideal op amp model. This is a simple three-terminal model, with two input terminals for the inverting and non-inverting op amp inputs, and one output terminal for the op amp output. We use bias point analysis to determine the output voltage in response to a 1 V input to this inverting amplifier configuration. The result of this analysis gives an output voltage of −10 V, as seen in Fig. 58, which is the expected output voltage.

EXAMPLE 9

Repeat the analysis of the circuit shown in Fig. 53, but this time use the μA741 model from the PSpice Eval library.

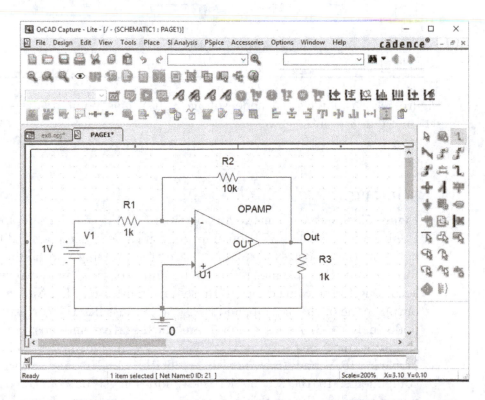

Figure 57: The schematic representation of the circuit in Fig. 53, using the PSpice model of an ideal op amp from the Analog library.

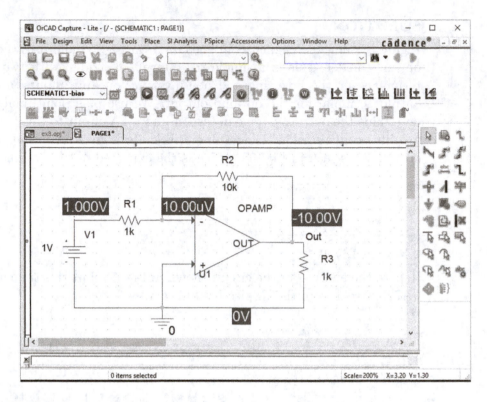

Figure 58: Displaying the node voltages after bias point analysis of the schematic in Fig. 57.

SOLUTION

Figure 59 shows a schematic for the circuit in Fig. 53 that employs the PSpice model of the μA741 op amp. As you can see, there are 7 different nodes in the model. Nodes 2 and 3 are the inverting and noninverting terminals of the op amp, respectively, and are wired in the schematic according to the connections to those terminals in Fig. 53. Node 6 is the output node of the op amp, and is also wired just like the corresponding node in Fig. 53. Nodes 4 and 7 are the negative and positive power supply connections, respectively. The corresponding nodes in Fig. 53, labeled -15 V and 15 V, are not explicitly wired to any other circuit components. In the schematic, these nodes must be connected to dc power supplies. To do this, we use off page connectors labeled VCC+

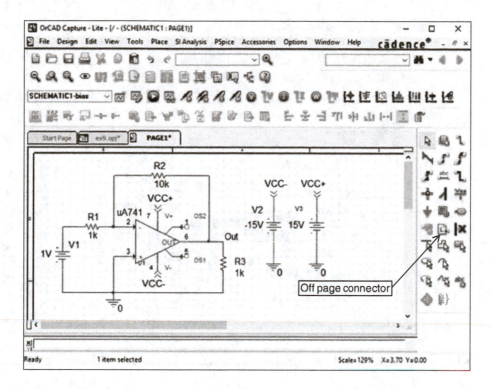

Figure 59: The schematic representation of the circuit in Fig. 53, using the PSpice model of the μA741 op amp.

and VCC$-$, and wire these labels to the appropriate dc power supplies to the right of the op amp circuit. This technique for drawing the schematic makes it less cluttered than it would be if we wired the dc power supplies directly to the op amp nodes. Finally, nodes 1 and 5 are not required to be connected to anything, so we leave them as is.

We use bias point analysis to simulate the response of this circuit to its 1 V dc input. The node voltages are displayed in Fig. 60, where we see that the output voltage is -9.998 V, close to the -10 V value predicted by the analysis in the previous section.

The model of the circuit for the μA741 component is quite complex. To see its details, highlight the μA741 component in the schematic shown in

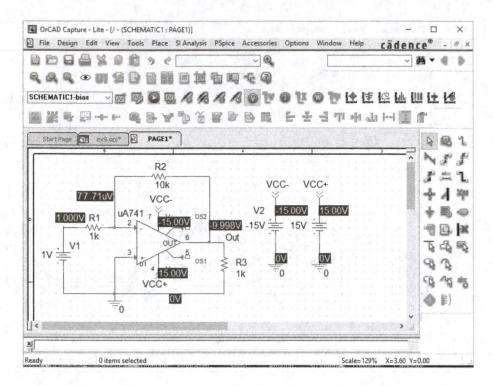

Figure 60: Displaying the node voltages after bias point analysis of the schematic in Fig. 59.

Fig. 59, right click the mouse and select Edit PSpice Model. The model's code, found in the Model Text window, is shown in Fig. 61. The model defines a sub-circuit with over thirty parts, including resistors, capacitors, and models of diodes and transistors. In ORCAD Capture you can change the component models to reflect the actual values of the sub-circuit components in a real device you are using, but PSpice Model Editor Lite only permits editing of diode models.

```
Model Text                                                                    [x]
*-----------------------------------------------------------------------------
* connections:   non-inverting input
*                | inverting input
*                | | positive power supply|
*                | | | negative power supply
*                | | | | output
*                | | | | |
.subckt uA741   1 2 3 4 5

c1    11 12 8.661E-12
c2     6  7 30.00E-12
dc     5 53 dy
de    54  5 dy
dlp   90 91 dx
dln   92 90 dx
dp     4  3 dx
egnd  99  0 poly(2),(3,0),(4,0) 0 .5 .5
fb     7 99 poly(5) vb vc ve vlp vln 0 10.61E6 -1E3 1E3 10E6 -10E6
ga     6  0 11 12 188.5E-6
gcm    0  6 10 99 5.961E-9
iee   10  4 dc 15.16E-6
hlim  90  0 vlim 1K
q1    11  2 13 qx
q2    12  1 14 qx
r2     6  9 100.0E3
rc1    3 11 5.305E3
rc2    3 12 5.305E3
re1   13 10 1.836E3
re2   14 10 1.836E3
ree   10 99 13.19E6
ro1    8  5 50
ro2    7 99 100
rp     3  4 18.16E3
vb     9  0 dc 0
vc     3 53 dc 1
ve    54  4 dc 1
vlim   7  8 dc 0
vlp   91  0 dc 40
vln    0 92 dc 40
.model dx D(Is=800.0E-18 Rs=1)
.model dy D(Is=800.00E-18 Rs=1m Cjo=10p)
.model qx NPN(Is=800.0E-18 Bf=93.75)
.ends
```

Figure 61: The Model Text window in the Model Editor for the μA741 op amp.

ASSESSMENT

♦ Use PSpice to solve the following problems: 5.5, 5.6, 5.17(a), 5.20(a), 5.31(a), and 5.42.

Chapter 6

INDUCTORS, CAPACITORS, AND NATURAL RESPONSE

PSpice parts that model inductors and capacitors are similar to those used for resistors. The primary property for these components is the inductance (in henries) or the capacitance (in farads). If the inductor has an initial current, it is specified in the PropertyEditor's IC property. If the capacitor has an initial voltage, it is also specified using the Property Editor's IC property.

6.1 TRANSIENT ANALYSIS

To examine a circuit's response as a function of time use transient analysis. Select Time Response (Transient) in the simulation settings dialog to select the parameters for transient analysis. You must specify the length of time for the simulation, in seconds, and the time to start storing data for the output. The simulation's duration depends on the circuit's time constants, so some preliminary circuit analysis is often necessary. The start time is usually zero, unless you want to collect data only after the circuit is in its steady-state.

You can also specify the step size, or time increment, to be used by PSpice, but it is not usually required. If you have used the Property Editor to specify initial conditions for any inductors or capacitors in your schematic, check the box that skips the initial bias point analysis of the circuit. We'll see these settings in the upcoming example.

You can examine the transient analysis output in two different ways. First, you can direct PSpice to output the transient analysis results to a file. Do this by clicking the Output File Options button in the transient analysis simulation settings dialog. This activates a second dialog, prompting you to specify the time increment used for the output data. Second, you can use Probe to plot voltages and currents of interest versus time.

6.2   NATURAL RESPONSE

To simulate a circuit's natural response, first find the initial inductor currents and capacitor voltages. Then create the schematic and edit the component properties to include those initial conditions. Example 10 illustrates the natural response of a series RLC circuit. We use a preliminary analysis to help us choose the simulation time settings and to check the simulation results.

EXAMPLE 10

a) Analyze the natural response of the circuit in Fig. 62 to find the type of damping, the peak value of v_c, and the frequency of oscillation.

b) Use PSpice to simulate the circuit's natural response. Plot the voltage v_c versus t, and use the Probe cursor to identify the peak value of the voltage.

SOLUTION

a) For the series RLC circuit shown in Fig. 62, $\alpha = R/2L = 1000$, $\alpha^2 = 10^6$, and $\omega_0^2 = 1/LC = 50 \times 10^6$. Hence the response is underdamped, and the roots of the characteristic equation are $s_1 = -1000 + j7000$ rad/s and $s_2 = -1000 - j7000$ rad/s. Therefore, capacitor voltage is

$$v_c = (B_1 \cos 7000t + B_2 \sin 7000t)e^{-1000t}.$$

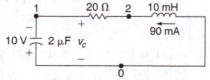

Figure 62: The circuit for Example 10.

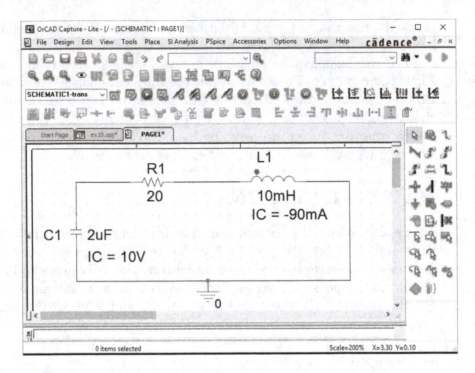

Figure 63: The schematic representation of the circuit in Fig. 62*.

From the initial conditions,

$$v_c(0) = -10 \quad \text{and} \quad \frac{dv_c(0)}{dt} = 45,000 \text{ V/s.}$$

Solving for B_1 and B_2 yields

$$v_c = (-10 \cos 7000t + 5 \sin 7000t)e^{-1000t} \text{ V,} \quad t \geq 0.$$

The peak value of v_c is 7.70 V, which occurs at $t = 362.29$ μs. The damped frequency is 7000 rad/s, and thus the damped period is 897.60 μs.

b) Figure 63 shows the PSpice schematic for the circuit in Fig. 62. We have set the initial values of capacitor voltage and inductor current using the Property Editors. Pay close attention to the orientation of these parts so the signs of the initial conditions are correct. The initial voltage is positive at the left-most terminal of the capacitor, or bottom terminal if the part is rotated once. Therefore, the initial capacitor voltage is

---

* In this schematic and any others containing inductors, we have edited the inductor part to eliminate the terminal labels 1 and 2, to simplify the schematic.

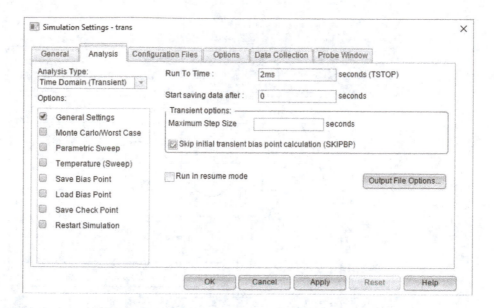

Figure 64: The simulation settings for transient analysis of the schematic in Fig. 63.

+10 V. The initial current enters the left-most terminal of the inductor, or bottom terminal if the part is rotated once. Therefore, the initial inductor current is −90 mA. These values are specified in the Property Editor column labeled IC.

From the preliminary analysis of the circuit, we set TSTOP = 2 ms in the Time Response simulation settings, to include the first two cycles of the response. We also check the box that skips the computation of the initial bias point. Figure 64 shows the completed simulation settings dialog.

The capacitor voltage plot is shown in Fig. 65. Display the cursor by selecting Trace/Cursor/Display, and then move the cursor manually with the right and left arrow keys, or select Trace/Cursor/Peak once the cursor is displayed. We see from the cursor output window that the peak output voltage is 7.7347 V, which occurs at $t = 371.222$ μs, which agrees with our analytic results.

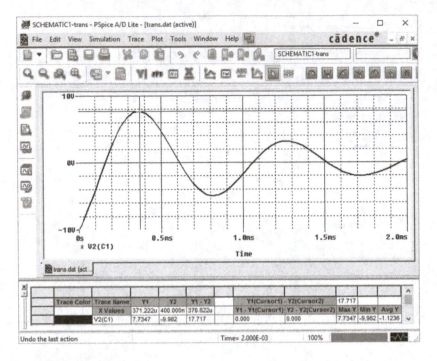

Figure 65: The Probe plot of the capacitor voltage for the circuit in Fig. 63, using the cursor to find the peak voltage.

ASSESSMENT

♦ Use PSpice to find and plot the indicated voltages and currents in the following problems: 8.2, 8.3, 8.4, 8.32, 8.33, and 8.34.

Chapter 7

THE STEP RESPONSE AND SWITCHES

We can find a circuit's step response by activating a dc source at the time the step occurs, usually at $t = 0$. This is illustrated in Example 11. Then, we present two of several ways to model switches in PSpice. The first involves a special voltage source, and the second uses a PSpice model of a realistic switch. We illustrate these two models of switches in Examples 12 and 13.

7.1   SIMPLE STEP RESPONSE

Example 11 simulates the step response of the circuit in Example 10 (Fig. 62) using a dc source.

EXAMPLE 11

a) The switch in the circuit shown in Fig. 66 is closed at $t = 0$. At the instant the switch is closed, the current in the inductor is 90 mA, and the voltage across the capacitor is 10 V, as shown in Fig. 66. Analyze

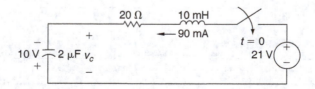

Figure 66: The circuit for Example 11.

the circuit to find its step response. Calculate the maximum value of v_c and the time it occurs.

b) Create a PSpice schematic, perform transient analysis, and plot v_c versus t from 0 to 4000 μs.

c) Compare the PSpice solution with the preliminary analysis.

SOLUTION

a) From Example 10 we know that the response is underdamped, that $s_1 = -1000 + j7000$ rad/s, $s_2 = -1000 - j7000$ rad/s, and the damped period is 897.60 μs. The step-response solution for v_c takes the form

$$v_c = v_f + (B'_1 \cos 7000t + B'_2 \sin 7000t)e^{-1000t}.$$

The initial values of v_c and dv_c/dt are the same as in Example 10, and the final value is 21 V. Hence $B'_1 = -31$ V and $B'_2 = 2$ V, so

$$v_c = 21 - (31 \cos 7000t - 2 \sin 7000t)e^{-1000t} \text{ V}, \quad t \geq 0.$$

The maximum value of v_c is 41.22 V at 419.32 μs.

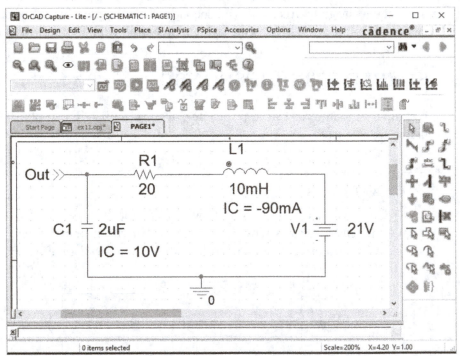

Figure 67: The schematic representation of the circuit in Fig. 66.

b) The PSpice schematic is shown in Fig. 67 and Fig. 68 shows the plot generated by Probe after transient analysis. We used both cursors, moving the second cursor by holding down the Shift key while using the left and right arrow keys. The cursors identify the values of $v_c(\max)$ and T_d directly on the Probe plot.

c) From the plot in Fig. 68,

$$v_c(\max) = 41.548 \text{ V}, \quad \text{at } 422.424 \ \mu s;$$

$$T_d = 961.682 \ \mu s;$$

$$v_f = 21 \text{ V}.$$

These results agree with the preliminary analysis.

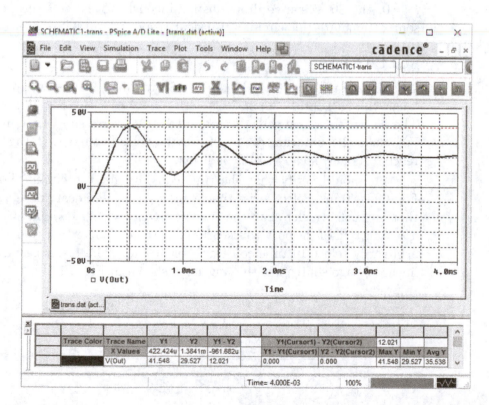

Figure 68: The Probe plot of the capacitor voltage for the schematic in Fig. 67.

7.2   PIECEWISE LINEAR SOURCES

A piecewise linear (PWL) time-dependent source allows you to define a single-valued function at a series of discrete times. PSpice determines the source's value at intermediate times using linear interpolation. Because the function must be single valued, the values of time used to define $f(t)$ must continually increase. One of the many uses for this type of source is to model a switch.

Example 12 uses a piecewise linear source to model an instantaneous drop in a source's voltage.

EXAMPLE 12

The circuit shown in Fig. 69 has been in operation for a long time. At $t = 0$, the 80 V source drops instantaneously to 20 V. Using a PWL source, simulate this circuit's transient response and plot $v_o(t)$ versus t.

SOLUTION

To simulate the drop in voltage from 80 to 20 V, we insert a piecewise linear voltage source of opposite polarity in series with an 80 V dc source. The piecewise source changes from 0 to 60 V over a short time interval. We make the duration of the 60 V step long enough that the circuit reaches its new steady-state condition.

You should verify that the time constant for the circuit shown in Fig. 69 is 40 ms. Select a rise time for the piecewise linear voltage source that is much smaller than the time constant, say 40 ms/1000 = 40 μs. The duration of the 60 V source should be much longer than 40 ms, so we arbitrarily select 10 time constants, or 400 ms. The resulting piecewise linear voltage source has the waveform shown in Fig. 70.

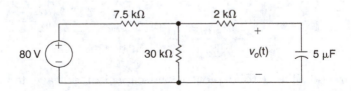

Figure 69: The circuit for Example 12.

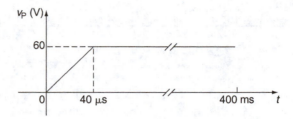

Figure 70: The waveform for the piecewise linear voltage source.

Figure 71 is the schematic for the circuit shown in Fig. 69. To specify the piecewise linear voltage source, enter pairs of time and voltage values in the appropriate columns of the Property Editor for this part. The first pair of values, with column headings T1 and V1, are both zero, since the value of the voltage source is 0 V at $t = 0$ s. The next pair of values, with column headings T2 and V2, are 40 μs and 60 V, since the value of the voltage source is 60 V at $t = 40$ μs. The final pair of values, with column headings T3 and V3, are 40 ms and 60 V, since the voltage source continues to supply 60 V for all time after $t = 40$ μs. Thus, the value

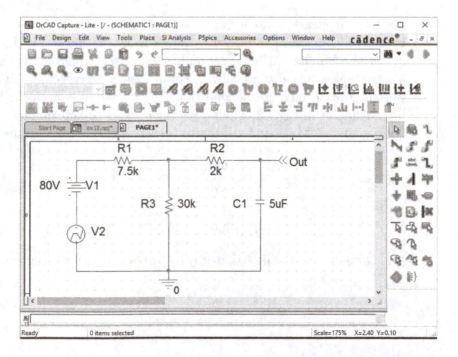

Figure 71: The schematic for the circuit in Fig. 69.

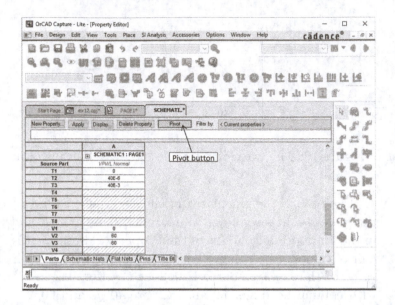

Figure 72: The Property Editor for the piecewise linear voltage source in Fig. 71. (Click the Pivot button to get this view.)

for T3 is arbitrary, so long as it is a time larger than T2. The completed Property Editor for the piecewise linear source is shown in Fig. 72.

In the Simulation Settings dialog we select transient analysis, but allow PSpice to calculate initial conditions automatically, since there were no prespecified initial conditions. Thus we don't check the SKIPBP box. The plot in Fig. 73, which uses the cursors to determine the output voltage at $t = 40$ ms and $t = 80$ ms, yields the following results:

t, ms	$v_o(t)$, V
0	64.0
40	33.714
80	22.488

You can see from the plot that $v_o(t)$ approaches 16 V as t approaches ∞.

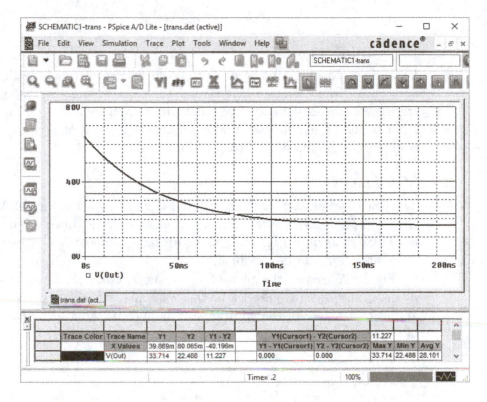

Figure 73: The Probe plot of the capacitor voltage for the schematic in Fig. 71.

7.3 REALISTIC SWITCHES

PSpice provides two special parts that represent actual switches. One is a normally-open switch, with the part name Sw_tOpen and the other is a normally-closed switch, with the part name Sw_tClose. Both are found in the Eval library. These parts include attributes of a realistic switch, including the equivalent resistance of the switch when open, the equivalent resistance of the switch when closed, the time when the switch opens (if it is a normally-closed switch) or closes (if it is a normally-open switch), and the transition time for switching from one state to the other.

We use the normally-open switch in Example 13 to examine the effect of nonzero switch resistance on the step response of an *RLC* circuit.

EXAMPLE 13

Use PSpice to simulate the circuit shown in Fig. 66. Include the voltage source from Example 11, and use the same values when specifying transient analysis. But assume when the switch is in the ON position (that is, the switch is closed) its resistance is 10 Ω. Plot v_c versus t and compare the results with those in Example 11.

SOLUTION

The PSpice schematic for the circuit in Fig. 66 is shown in Fig. 74. We use the Property Editor for the switch to specify its resistance, when closed, is 10 Ω, as shown in Fig. 75. This figure shows some other properties of the switch that can be modified.

Figure 76 shows the plot of v_c versus t. Note that the values of $v_c(\text{max})$ and T_d are different in this example, when compared to Example 11, because of the additional resistance introduced by the model of the switch.

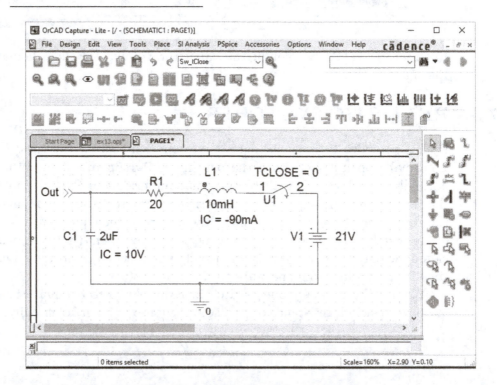

Figure 74: The schematic for the circuit in Fig. 66, using a normally-open switch part.

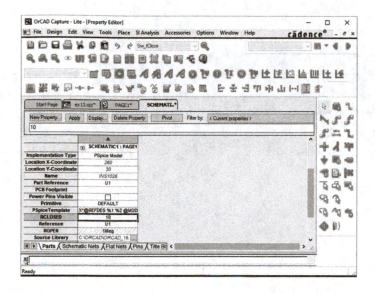

Figure 75: The Property Editor for the normally-open switch in the schematic of Fig. 74.

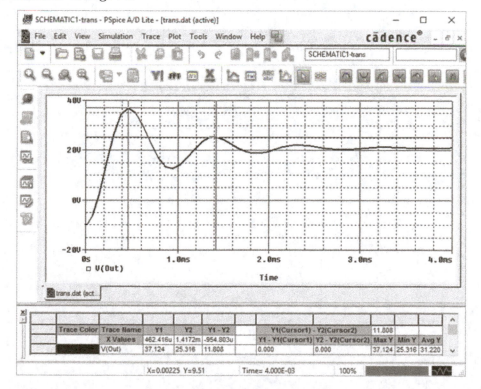

Figure 76: The Probe plot of capacitor voltage for the schematic in Fig. 74.

ASSESSMENT

♦ Use PSpice to find and plot the indicated voltages and currents in the
following problems: 7.5, 7.14, 7.22(a) and (b), 7.29, 7.46, 7.65, 8.29,
8.49, and 8.63.

Chapter 8

VARYING COMPONENT VALUES

In PSpice, you can use a global parameter as the value of a circuit component. You can specify that the simulation vary the value of the global parameter, thereby varying the value of the circuit component. You can thus explore the effects of changing a circuit component's value on the behavior of the circuit.

Example 14 demonstrates how easily PSpice can simulate the same circuit several times with a different value of a given component used in each simulation. We use Probe to plot the circuit's output for each component value.

EXAMPLE 14

Use a global parameter to vary the value of the resistor in Fig. 62 (see Example 10) from 20 Ω to 100 Ω in 20 Ω steps. Then plot v_c versus t for each of the resistor values.

SOLUTION

The schematic for the circuit in Fig. 62 is shown in Fig. 77. The capacitor and the inductor are modeled just as they were in Example 10, but the resistor model is different: a global parameter, RESISTANCE, has replaced the resistor's value. We define the global parameter by adding

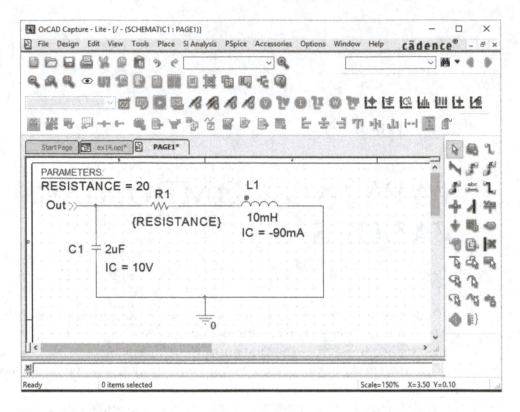

Figure 77: The schematic for the circuit in Fig. 62 with a variable resistance.

the Param part from the Special library to the circuit schematic, as can be seen in Fig. 77.

Use the Param part's Property Editor to define the global parameter. To do this, open the Property Editor, click the New button, and type the name of the global parameter in the dialog box. This creates a new property, with the name RESISTANCE, for the Param component. Enter a default value of 20 Ω for this new property. To have this property's name and its value appear in the schematic, highlight the spreadsheet cell with the new property and click the display button. Then choose Display Name and Value from the Display Format dialog. The completed Property Editor for the Param part is shown in Fig. 78.

Next specify the analysis using the Simulation Settings window. Select Time Domain, or transient, analysis, with the same final time as we have used previously with this circuit, namely 2 ms. But now we also

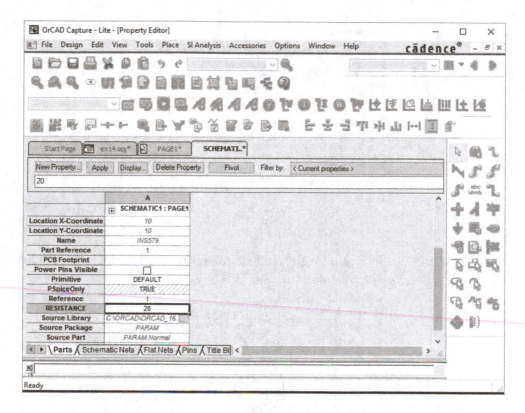

Figure 78: The Property Editor for the Param part, after adding the RESISTANCE property.

check the box labeled Parametric Sweep. As shown in Fig. 79, we then select a Global parameter as the sweep variable, identify the parameter name, and input the start value, end value and increment for the sweep variable.

When we run the simulation, PSpice actually simulates transient analysis for five different circuits, one circuit for each of the five different resistance values we specified. Select all five of these results to plot, as shown in Fig. 80. Note that the circuit with the smallest value of resistance has the smallest damping ratio. Increasing the value of resistance increases the damping ratio. You should confirm these results analytically.

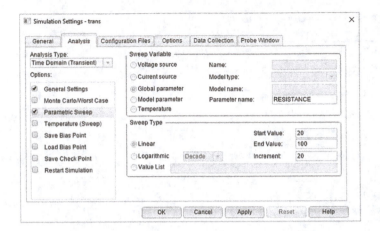

Figure 79: Specifying a parametric sweep of the RESISTANCE global parameter for the schematic in Fig. 77.

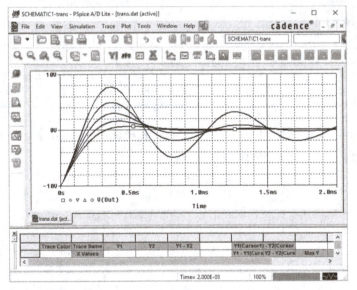

Figure 80: The Probe plot of the capacitor voltage in the circuit of Fig. 62 as the resistance is varied from 20 Ω to 100 Ω in 20 Ω increments. The plot with the highest peak corresponds to the circuit whose resistance is 20 Ω.

ASSESSMENT

♦ Use PSpice to solve the following problems: 4.69, 4.87, 4.91, 8.17, 8.18, 8.19, and 8.39.

Chapter 9

SINUSOIDAL STEADY-STATE ANALYSIS

Now we use PSpice to examine a circuit's ac steady-state response. This is the response to a sinusoidal input at a fixed frequency, once the transient response has decayed to zero.

9.1 SINUSOIDAL SOURCES

The following equations describe the damped sinusoidal voltage source:

$$v_g = \text{VOFF} + \text{VAMPL} \sin [\text{PHASE}], \quad 0 \le t \le \text{TD};$$
$$v_g = \text{VOFF} + \text{VAMPL} e^{-\text{DF}(t-\text{TD})} \sin[2\pi\text{FREQ}(t - \text{TD}) + \text{PHASE}],$$
$$\text{TD} \le t \le \text{TSTOP};$$

where

$$
\begin{aligned}
\text{VOFF} &= \text{offset voltage in volts;} \\
\text{VAMPL} &= \text{amplitude in volts;} \\
\text{FREQ} &= \text{frequency in hertz;} \\
\text{TD} &= \text{delay in seconds;} \\
\text{DF} &= \text{damping factor in seconds}^{-1}; \text{ and} \\
\text{PHASE} &= \text{phase angle in degrees.}
\end{aligned}
$$

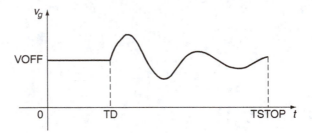

Figure 81: A damped sinusoidal voltage.

The default values of TD and DF are zero, and FREQ defaults to 1/TSTOP. Figure 81 shows a graph of a damped sinusoidal voltage.

Before using a time-dependent source in a circuit schematic, you should check a plot of its waveform. To do this, simulate the source in a purely resistive circuit. The output waveform in a resistive circuit is a scaled replica of the input waveform. You can choose a convenient scaling factor for the test circuit. For example, assume you want to check a damped sinusoidal voltage source described as

$$v_g = 12.5 + 25e^{-0.1(t-1)}\,\sin[0.4\pi(t-1)]\;\mathrm{V}.$$

If this is the voltage in the voltage-divider circuit shown in Fig. 82, the output voltage is $0.8v_g$.

The part name for the sinusoidal voltage source is Vsin and is found in the Source library. We specify the values for TD, DF, VOFF, VAMPL, FREQ, and PHASE using the Property Editor, as shown in Fig. 83. The output voltage for the transient analysis of Fig. 82 is shown in Fig. 84, and confirms that we have correctly specified the properties of the sinusoidal voltage source.

9.2   SINUSOIDAL STEADY-STATE RESPONSE

The sinusoidal sources discussed in the previous section are usually used to simulate a circuit's transient response. If we want to simulate a circuit's sinusoidal steady-state response (also known as the ac steady-state response), we use ac voltage or current sources, whose part names are Vac and Iac, respectively. Then, AC Sweep analysis calculates the output voltage or current phasors for one frequency or over a range of different frequency values. Example 15 simulates the single-frequency steady-state sinusoidal response using ac sources and AC Sweep analysis.

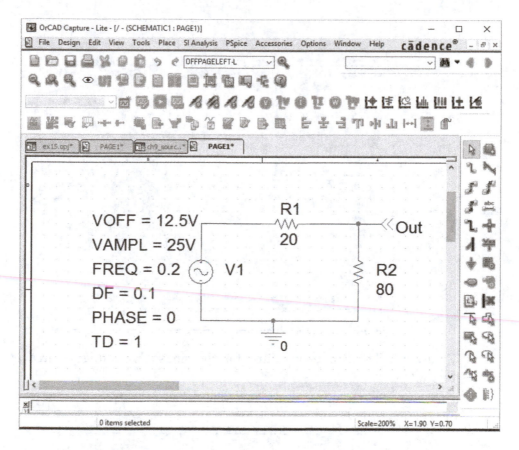

Figure 82: A voltage divider schematic, used to check the properties of a sinusoidal voltage source model.

EXAMPLE 15

The sinusoidal sources in the circuit shown in Fig. 85 are described by the equations

$$v_g = 20 \cos(10^5 t + 90°) \text{ V} \quad \text{and} \quad i_g = 6 \cos 10^5 t \text{ A}.$$

a) Use PSpice to find the magnitude and phase of v_o and i_o.
b) Compare the PSpice results in (a) to an analytic solution for v_o and i_o.

SOLUTION

a) The schematic for the circuit in Fig. 85 is shown in Fig. 86. We specify the magnitudes and phase angles of the ac sources using their Property Editors. Note that the phase angle property assumes that the source is

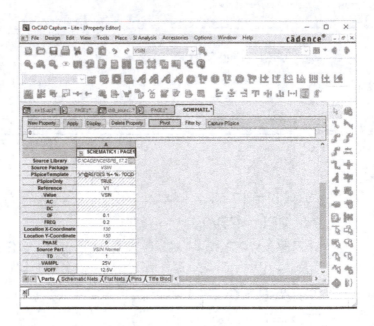

Figure 83: The Property Editor for the sinusoidal voltage source part in Fig. 82.

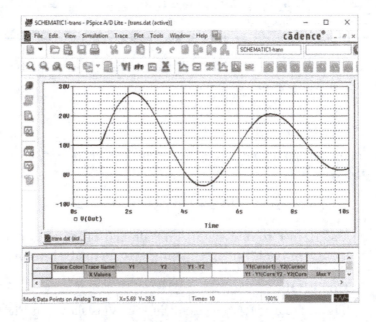

Figure 84: The output voltage plot after transient analysis of the schematic in Fig. 82.

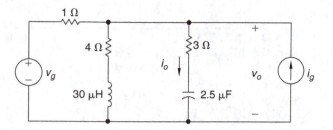

Figure 85: The circuit for Example 15.

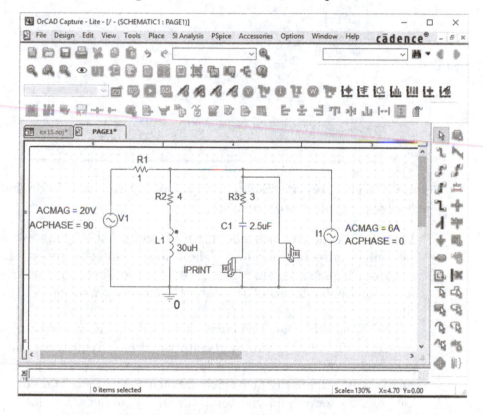

Figure 86: The schematic for the circuit in Fig. 85.

specified as a cosine, and the phase angle units are degrees. The Property Editor for the ac voltage source is shown in Fig. 87.

Note that the schematic in Fig. 86 includes current and voltage printers that will print the output voltage and current. Using the Property Editors for the printers, specify the output from AC analysis by placing a "Y" (for "yes") in the column labeled AC. Place a Y in the columns labeled MAG and PHASE if you want the output values in polar form.

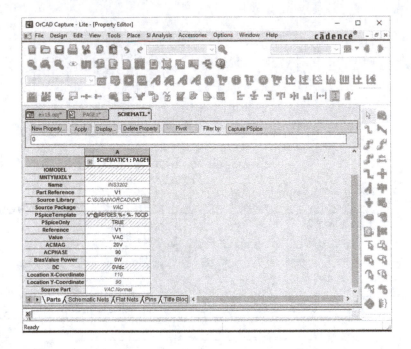

Figure 87: The Property Editor used to specify the magnitude and phase angle of the ac voltage source in Fig. 86.

Place a Y in the columns labeled REAL and IMAG if you want the output values in rectangular form. The voltage printer's Property Editor is shown in Fig. 88. As you can see, we have specified that both the polar and rectangular forms of the output voltage be written to the output file.

Next, specify the Simulation Settings for AC Sweep analysis, as shown in Fig. 89. We supply the start and end frequencies and the total number of frequency points for the sweep. Since we want PSpice to analyze the schematic for a single frequency, the start and end frequencies are the same and the total number of frequency values is 1. All frequencies are specified in hertz.

Part of the output file is shown in Fig. 90. The PSpice simulation results are $\mathbf{V_o} = 16.31\underline{/71.51°}$ V   and $\mathbf{I_o} = 3.261\underline{/124.6°}$ A.

b) To find the phasor voltage $\mathbf{V_o}$, solve a single node-voltage equation:

$$\frac{\mathbf{V_o}}{4+j3} + \frac{\mathbf{V_o}}{3-j4} + \frac{\mathbf{V_o}-j20}{1} = 6\underline{/0°}.$$

The solution for $\mathbf{V_o}$ is

$$\mathbf{V_o} = 16.31\underline{/71.51°} \text{ V}.$$

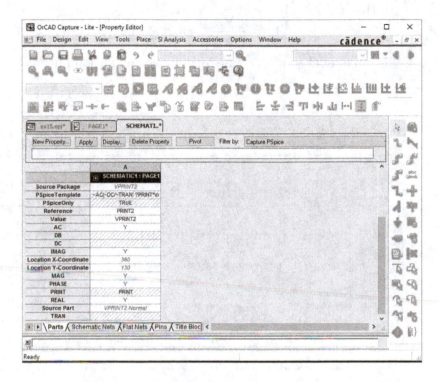

Figure 88: The Property Editor for the voltage printer in Fig. 86.

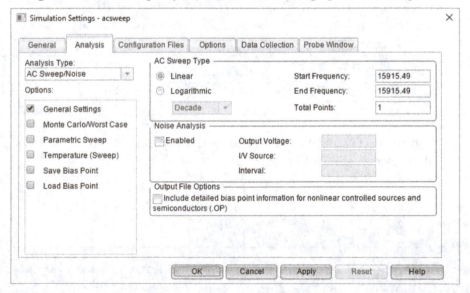

Figure 89: The Simulation Settings dialog for AC Sweep analysis of the schematic in Fig. 86.

Hence

$$\mathbf{I}_o = \frac{\mathbf{V}_o}{3 - j4} = 3.26\underline{/124.64°} \text{ A}.$$

The analytical solutions thus match the simulation output.

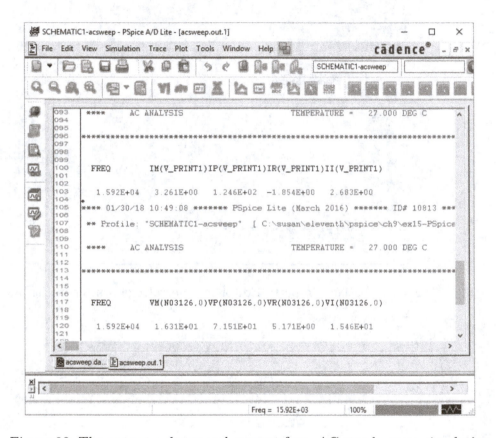

Figure 90: The output voltage and current from AC steady-state simulation of the schematic in Fig. 86.

ASSESSMENT

♦ Use PSpice to solve the following problems: 9.28, 9.33(a), 9.40, 9.54, 9.71, and 9.72.

Chapter 10

LINEAR AND IDEAL TRANSFORMERS

PSpice supports simulation of three types of mutually-coupled coils.

- A mutually-coupled coil can be modeled using two inductors and a part named K from the Analog library, which is used to select the coupling coefficient for the inductors.

- Linear transformers are modeled with the **XFRM_Linear** part from the Analog library. The properties specified for this part include the inductance of the two coils and the coefficient of coupling for the transformer.

- Ideal transformers are also modeled with the **XFRM_Linear** part from the Analog library. The two inductance values and the coefficient of coupling are specified to meet the requirements of an ideal transformer.

This chapter has examples of both linear and ideal transformers.

10.1 LINEAR TRANSFORMERS

The PSpice part that models a linear transformer is named **XFRM_Linear** and is found in the Analog library. We use this part in Example 16.

EXAMPLE 16

a) Use PSpice to find the rms amplitudes and the phase angles of i_1 and i_2 in the circuit shown in Fig. 91 when $v_g = 141.42 \cos 10t$ V.

77

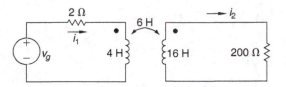

Figure 91: The circuit for Example 16.

b) Check the simulation results by using them to show that the average power generated equals the average power dissipated.

SOLUTION

a) Begin by noting that $V_g = 100\underline{/0°}$ V(rms), $f = 5/\pi \approx 1.59155$ Hz, and the coefficient of coupling $k = 6/\sqrt{64} = 0.75$. Figure 92 shows the circuit schematic. Note that both the primary and the secondary sides of the linear transformer must be connected to the zero node, or ground.

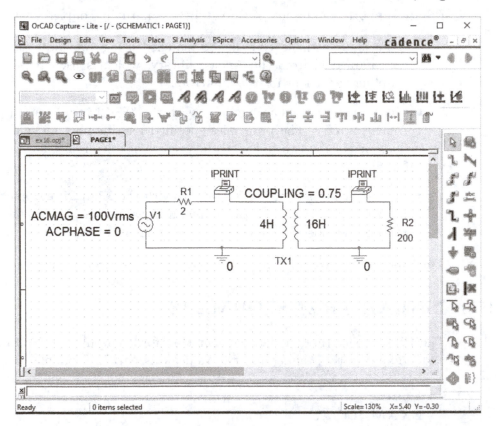

Figure 92: The schematic for the circuit in Fig. 91.

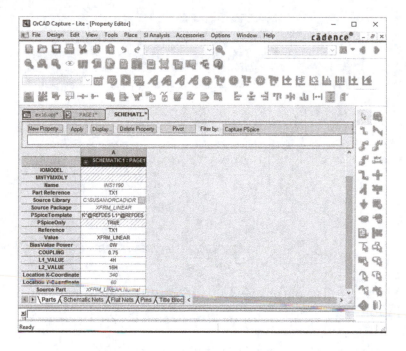

Figure 93: The Property Editor for the linear transformer part in Fig. 92.

We have inserted current printers in the schematic to include the current values from AC Sweep simulation in the output file.

Specify the inductance values for the primary and secondary coils and the coupling coefficient value using the Property Editor for the linear transformer part, as shown in Fig. 93. These properties are also displayed in the schematic of Fig. 92. Specify AC Sweep analysis at the single frequency 1.59155 Hz.

The PSpice output file is shown in Fig. 94. From this output file,

$$\mathbf{I}_1 = 2.958\underline{/-67.43°} \text{ A(rms)};$$
$$\mathbf{I}_2 = 0.6929\underline{/-16.09°} \text{ A(rms)}.$$

b) Using the PSpice solution, we find that the total average power generated is

$$P_g = (100)(2.958) \cos 67.43° = 113.532 \text{ W}.$$

Figure 94: The output file from AC steady-state simulation of the schematic in Fig. 92.

The average power dissipated in R_1 is

$$P_1 = (2.958)^2(2) = 17.500 \text{ W}.$$

The average power dissipated in R_2 is

$$P_2 = (0.6929)^2(200) = 96.022 \text{ W}.$$

Because $P_g = P_1 + P_2$, the PSpice solution is consistent with the conservation-of-energy principle.

10.2 IDEAL TRANSFORMERS

You can use the XFRM_Linear part to model an ideal transformer by making L_1 and L_2 large enough to make ωL_1 and ωL_2 much greater than the other impedances in the circuit. You must also set the coefficient of coupling to 1. Specify the turns ratio of the ideal transformer in the model using the relationship $L_1/L_2 = (N_1/N_2)^2$. Example 17 illustrates these requirements.

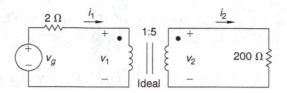

Figure 95: The circuit for Example 17.

EXAMPLE 17

Find the magnitude and phase angle of the primary and secondary voltages for the circuit in Fig. 95, where $v_g = 50 \cos 1000t$ V.

SOLUTION

From the circuit diagram, $N_1/N_2 = 1/5$ so $L_2 = 25L_1$. In picking values for L_1 and L_2, we want $\omega L_2 \gg 200$ Ω. Since $\omega = 1000$ rad/s, if we make $L_2 = 200$ H, ωL_2 will be three orders of magnitude greater than R_2. Therefore, choose $L_2 = 200$ H and $L_1 = 8$ H.

We represented the circuit in Fig. 95 as a schematic in Fig. 96. Note that both sides of the transformer are connected to ground, and that we have inserted printers to write the node voltages at the printer locations to the output file after simulation. As before, we specified AC Sweep

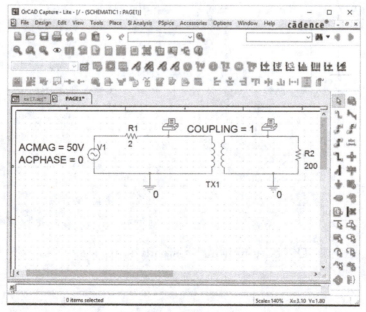

Figure 96: The schematic representation of the circuit in Fig. 95.

analysis at a single frequency of 159.1549 Hz. The pertinent portion of the output file is shown in Fig. 97. Hence

$$\mathbf{V}_1 = 40\underline{/0.01146°}\ \text{V} \quad \text{and} \quad \mathbf{V}_2 = 200\underline{/0.01146°}\ \text{V}.$$

You should verify that the analytic solution for $\mathbf{V}_1$ and $\mathbf{V}_2$ yields

$$\mathbf{V}_1 = 40\underline{/0°}\ \text{V} \quad \text{and} \quad \mathbf{V}_2 = 200\underline{/0°}\ \text{V}.$$

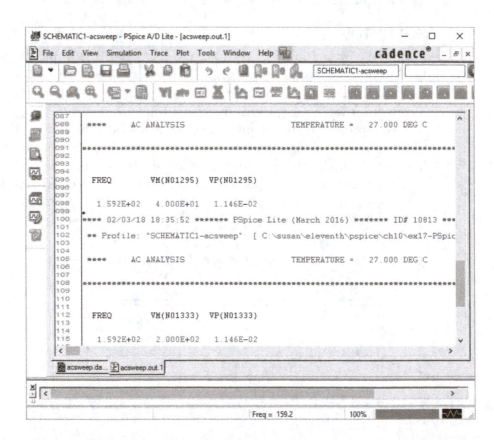

Figure 97: The output from ac steady-state simulation of the schematic in Fig. 96.

ASSESSMENT

♦ Use PSpice to solve the following problems: 9.74(a), 10.36(a), 10.37(c), and 10.56(a).

Chapter 11

COMPUTING AC POWER WITH PROBE

In this chapter we investigate the relationships among current, voltage, instantaneous power, average power, and reactive power using Probe. Although we introduce several new Probe features, you do not need any new PSpice constructs in order to investigate ac power.

To begin, construct a schematic for the parallel RC circuit in Fig. 98, as shown in Fig. 99. Simulate the circuit using transient analysis for 100 μs. When the blank graph appears, select the menu item Plot/Add Plot to Window three times to create four blank plots on the screen. Then select the top plot on the screen, and plot the voltage v_1. The plot should look like the first one (top) shown in Fig. 100.

Next, select the second plot on the screen, and plot the current through the source, i_s. Note the phase relationship between the source voltage and current: The current leads the voltage because the load is capacitive.

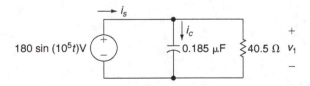

Figure 98: A circuit used to investigate AC power relationships.

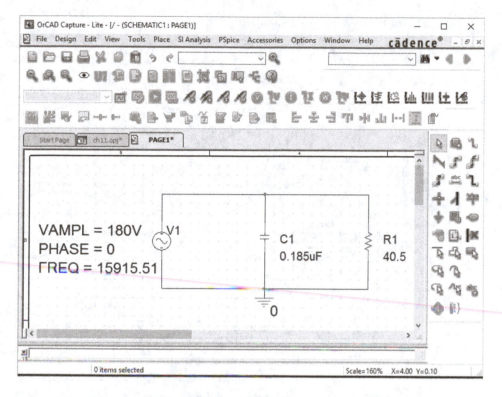

Figure 99: The schematic for the circuit in Fig. 98.

Compute the power factor angle of the load as follows:

$$
\begin{aligned}
Z_{\mathrm{L}} &= R \| (1/j\omega C) \\
&= \frac{R}{j\omega RC + 1}. \\
\underline{/Z_{\mathrm{L}}} &= 0 - \arctan(\omega RC) \\
&= -\arctan(10^5 \cdot 40.5 \cdot 0.185 \times 10^{-6}) \\
&= -36.84°.
\end{aligned}
$$

Now, select the third plot, and graph the instantaneous power supplied by the source. Recall that the expression for instantaneous power is $v(t) \cdot i(t)$. But Probe has a variable that represents the instantaneous power of every component, so just select $-W(V1)$. The minus sign $(-)$ inverts the sign of the instantaneous power to conform to the PSpice sign convention. In this plot, you have confirmed that the frequency of the instantaneous power is double the frequency of the voltage and current.

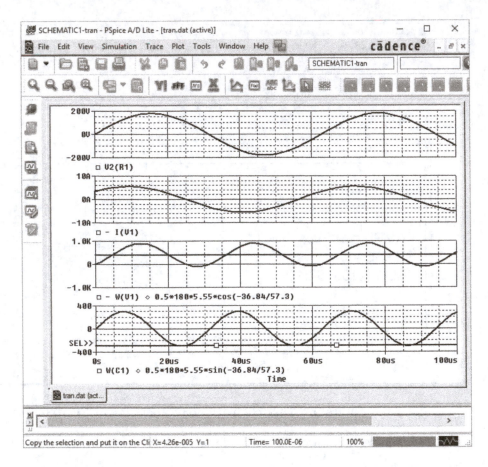

Figure 100: Plots of some power quantities for the circuit in Fig. 98: Top plot, source voltage; Second plot, source current; Third plot, instantaneous and average power supplied by the source; Bottom plot, instantaneous and reactive power for the capacitor.

Add a plot of the average power supplied by the source to the third plot. The formula for average power is

$$P = \frac{1}{2}V_{\mathrm{m}}I_{\mathrm{m}} \ \cos(\underline{/Z_{\mathrm{L}}}).$$

We need the magnitudes of the source voltage and its current so select the top two plots, one at a time, and use the cursor to find the maximum values $V_{\mathrm{m}} = 180$ and $I_{\mathrm{m}} = 5.55$. Select the third plot and add the trace described by the expression

$$.5 * 180 * 5.55 * \cos(-36.84/57.3),$$

using the menu item Trace/Add Trace and typing the expression into the
Trace Expression box. Note that we divided the impedance angle ($\underline{/Z_L}$ =
$-36.84°$) by 57.3 to convert the angle from degrees to radians, as required
by Probe. The third plot in Fig. 100 shows the average power is a straight
line centered on the plot of instantaneous power, as expected. Use the cursor
to find the average power is 400 W.

Finally, we examine the reactive power in the circuit. On the fourth
plot, add a trace of the instantaneous power absorbed by the capacitor,
using W(C1). Note that this instantaneous power has an average value of
zero. Add a trace of the reactive power supplied by the source to this plot,
which is described by the expression

$$.5 * 180 * 5.55 * \sin(-36.84/57.3).$$

You have completed the plot shown in Fig. 100. Now try to describe the
relationship between the instantaneous and reactive power for the capacitor.

We have briefly explored the ac power characteristics of the circuit shown
in Fig. 98 using Probe. You should take the time to investigate further; try
to construct plots of complex power, for example.

ASSESSMENT

♦ Use PSpice to solve the following problems: 10.7, 10.18(a), and 10.65(a).

Chapter 12

FREQUENCY RESPONSE

We now examine the features in PSpice that simulate a circuit as its input frequency is varied, allowing us to characterize the frequency response of the circuit.

12.1 SPECIFYING FREQUENCY VARIATION AND NUMBER

We use AC Sweep analysis, introduced in Section 9.2 for sinusoidal steady-state analysis, to simulate the frequency response of a circuit. You can vary the frequency linearly or logarithmically. If you choose logarithmic variation, you must then specify decade or octave. Recall that a decade is a 10-to-1 change in frequency and that an octave is a 2-to-1 change in frequency. After selecting the type of frequency variation, you then specify the number of frequencies (points) if the variation is linear or the number of frequencies (points) per decade (octave) if the frequency variation is decade (octave). As before, you must also input start and end frequencies for the sweep, using the units Hz.

If you select a linear variation, the frequency increment Δf is

$$\Delta f = \frac{\text{FSTOP} - \text{FSTART}}{\text{NP} - 1}.$$

For example, if FSTOP = 1801 Hz, FSTART = 1 Hz, and the number of points is 181, then

$$\Delta f = \frac{1801 - 1}{181 - 1} = 10 \text{ Hz}.$$

If you select a decade variation, the frequencies within a given decade are

$$f_k = \text{FSD} \cdot 10^{k/\text{ND}}$$

where FSD is the frequency at the start of the decade, ND is the number of points per decade, and k is an integer in the range from 1 to ND. For example, if ND = 20, FSTART = 50, and FSTOP = 5000, the frequencies in the decade between 50 and 500 would be

$$f_k = (50)10^{k/20}$$

where $k = 1, 2, 3, \ldots, 20$. Hence

$$f_1 = (50)10^{1/20} = 56.10 \text{ Hz};$$
$$f_2 = (50)10^{2/20} = 62.95 \text{ Hz};$$
$$\vdots$$

If you select an octave variation, the frequencies within a given octave are

$$f_k = \text{FSO} \cdot 2^{k/\text{NO}}$$

where FSO is the frequency at the start of the octave, NO is the number of points per octave, and k is an integer in the range from 1 to NO. For example, if FSTART = 400 Hz, FSTOP = 3200 Hz, and NO = 10, the frequencies in the octave from 400 to 800 Hz would be

$$f_k = (400)2^{k/10}.$$

Hence

$$f_1 = (400)2^{0.1} = 428.71 \text{ Hz};$$
$$f_2 = (400)2^{0.2} = 459.48 \text{ Hz};$$
$$\vdots$$

12.2 FREQUENCY RESPONSE OUTPUT

Probe plots are typically used to output AC Sweep simulation results. Printer parts inserted into the schematic can be used to write simulation results to the output file. You can specify voltage and current values in rectangular, polar, or decibel form by placing a Y in the appropriate column for the printer's Property Editor.

Example 18 simulates the frequency response of a parallel *RLC* circuit using PSpice.

EXAMPLE 18

a) The current source in the circuit shown in Fig. 101 is 50 cos ωt mA. Use Probe to plot v_o versus f from 1000 to 2000 Hz in increments of 10 Hz on a linear frequency scale.

b) From the Probe plot, estimate the center frequency, the bandwidth, and the quality factor of the circuit.

c) Compare the results obtained in (b) with an analytic solution for f_0, β, and Q.

SOLUTION

a) The PSpice schematic for the circuit in Fig. 101 is shown in Fig. 102. We specify the parameters of the AC Sweep as shown in Fig. 103.

b) Figure 104 shows the Probe plot of v_o versus f, using the linear range of frequencies shown in Fig. 104. Using the Probe Cursor, we find the peak amplitude of about 400 V (399.697 V) occurs at a frequency of 1590 Hz in Fig. 104(a). Thus we estimate the center frequency at 1590 Hz.

To estimate the bandwidth, use both cursors to find the frequencies where $v_o = 400/\sqrt{2} = 282.84$ V. From the Probe plot in Fig. 104(b), the closest values are 282.759 V at 1552.1 Hz and 282.759 V at 1632.0 Hz. The bandwidth is then shown as the variable "Y1 – Y2" in the Probe Cursor Window below the plot, and has the value 79.824, or about 80 Hz.

Calculate the quality factor from the relationship

$$Q = f_0/\beta = 1590/80 = 19.9.$$

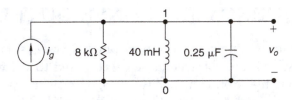

Figure 101: The circuit for Example 18.

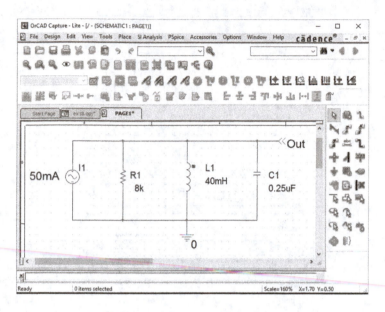

Figure 102: The schematic for the circuit in Fig. 101.

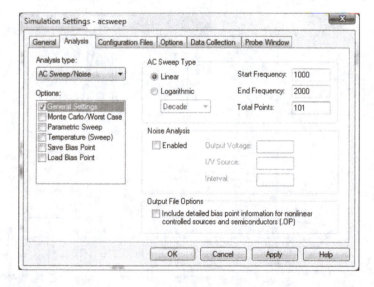

Figure 103: Specifying the parameters for the AC Sweep of the schematic in Fig. 102.

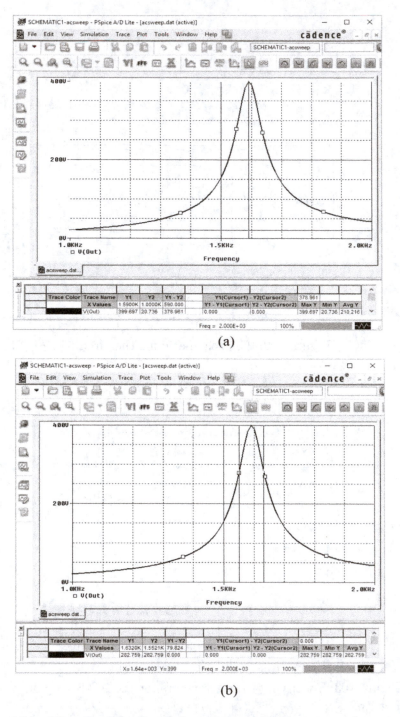

Figure 104: The Probe plot of output voltage versus frequency for Example 18. The cursors identify (a) the center frequency and (b) the bandwidth.

c) A direct analysis of the circuit yields

$$f_0 = 1591.55 \text{ Hz};$$
$$f_1 = 1552.26 \text{ Hz};$$
$$f_2 = 1631.84 \text{ Hz};$$
$$\beta = f_2 - f_1 = 79.58 \text{ Hz};$$
$$Q = f_0/\beta = 20.$$

We summarize the comparison with the PSpice analysis as follows:

Quantity	Analysis	PSpice
f_0 (Hz)	1591.55	1590.0
f_1 (Hz)	1552.26	1552.2
f_2 (Hz)	1631.84	1631.9
β (Hz)	70.58	79.8
Q	20.00	19.9

If necessary, we can improve the accuracy around the center frequency by rerunning the simulation using different values for the Start and Stop frequencies and a larger number of points.

In Example 19, we use a global parameter to vary the value of the capacitor in the circuit shown in Fig. 101. We use Probe to look at how the capacitor's value affects the frequency response characteristics of the circuit.

EXAMPLE 19

Modify the PSpice schematic for Example 18 to step the capacitor values from 0.15 μF through 0.35 μF in increments of 0.5 μF. Then plot the frequency response for all values of capacitance using Probe. Comment on the effect of the changing capacitance.

SOLUTION

The modified schematic is shown in Fig. 105. We have replaced the single-valued capacitor with a capacitor whose value is {CVAL}. We have also added a Param part with a new attribute called CVAL.

We have also made two changes to the Simulation Settings. First, we changed the AC Sweep to start at 500 Hz and end at 2500 Hz, to

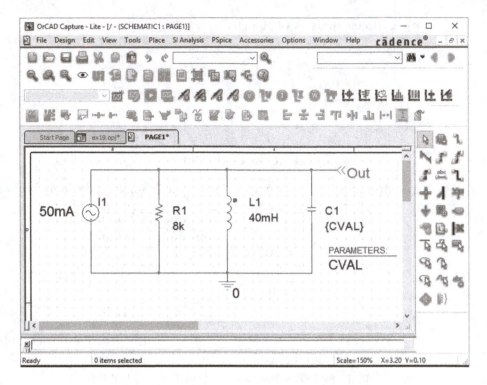

Figure 105: The schematic for the circuit in Fig. 101.

accommodate the broader frequency band that results from the range of capacitor values. Second, we have added a Parametric Sweep, to vary the global parameter CVAL from 0.15 μF to 0.35 μF in steps of 0.05 μF.

Figure 106 shows the Probe plot. The smallest value of capacitance produced the plot farthest to the right. As the capacitance increases, the plots move to the left. Hence the center frequency *decreases* as the capacitance *increases*. But we expect this result because the equation for center frequency of an RLC circuit is

$$\omega_0 = \sqrt{\frac{1}{LC}}.$$

Furthermore, as the capacitance increases, the center peak becomes sharper, as you can see in Fig. 106. That is, as the capacitance increases, the quality also increases. This result, too, comes as no surprise because the equation for Q in a parallel RLC circuit is

$$Q = R\sqrt{\frac{C}{L}}.$$

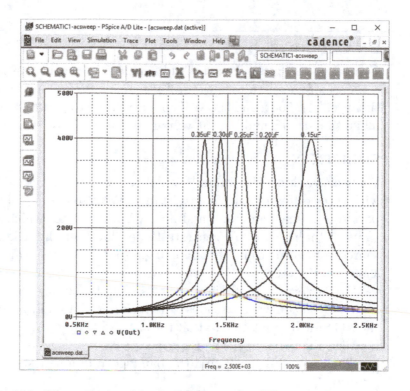

Figure 106: The relationship between a circuit's capacitance and its frequency response.

12.3 BODE PLOTS WITH PROBE

In Section 12.2, we generated a frequency response plot for a circuit, using PSpice and Probe. The voltage and current PRINT parts for frequency response data provides output voltages and currents in several different forms: real and imaginary parts, magnitude and phase angles, and dB magnitude and phase angles. Probe supports plotting frequency response data using these same forms. In Section 12.2, we plotted linear voltage magnitude versus linear frequency. We now turn to another form: dB voltage magnitude versus log frequency and phase angle versus log frequency. This is the form used for Bode plots.

Example 20 explores the frequency response of a series RLC circuit as the resistance is varied. This example presents the circuit's frequency response in Bode plot form.

EXAMPLE 20

Construct a PSpice schematic and simulate it to generate the frequency response of the circuit shown in Fig. 107 for three different values of resistance: 5 Ω, 50 Ω, and 500 Ω. Then use Probe to plot the output voltage magnitude in dB and output voltage phase angle versus log frequency.

SOLUTION

We use a resistor with value {RVAL} to simulate the circuit three times with three different values of resistance. We also add a global parameter part to our schematic and add an RVAL attribute, which becomes the value of the resistor. To generate Bode plots, set the magnitude of the source voltage to 1 and its phase angle to 0°. The resulting schematic is shown in Fig. 108.

We perform a preliminary analysis of the circuit to find its center frequency. Then we can choose the starting and ending frequencies for the AC Sweep simulation so that the circuit's center frequency is centered between the two end-point frequencies.

The center frequency and the bandwidth of a series RLC circuit are given by

$$\omega_0 = \sqrt{\frac{1}{LC}} \quad \text{and} \quad \beta = \frac{R}{L}.$$

For the circuit shown in Fig. 107, the center frequency is

$$\omega_0 = \frac{1}{\sqrt{(25 \times 10^{-3})(1 \times 10^{-6})}} = 6324.56 \text{ rad/sec}$$
$$\approx 1000 \text{ Hz},$$

and the bandwidth ranges from

$$\beta_{\text{narrow}} = \frac{5}{25 \times 10^{-3}} = 200 \text{ rad/sec}$$
$$\approx 3.2 \text{ Hz}.$$

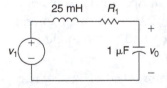

Figure 107: The circuit for Example 20.

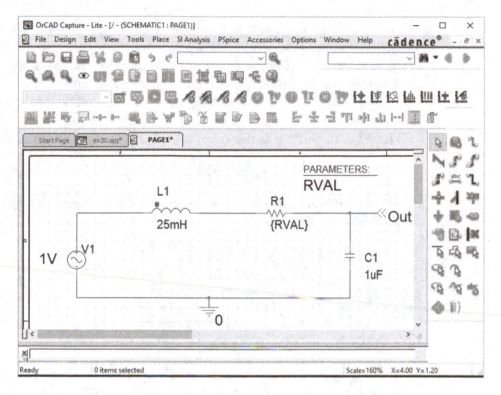

Figure 108: The schematic for the circuit in Fig. 107.

to

$$\beta_{\text{wide}} = \frac{500}{25 \times 10^{-3}} = 20,000 \text{ rad/sec}$$

$$\approx 32 \text{ kHz}.$$

Thus we choose logarithmic frequency variation with 100 Hz as the starting frequency and 10,000 Hz as the ending frequency. This places the center frequency in the middle of the logarithmic frequency axis and provides an adequate frequency range for the wideband circuit's response.

After entering these values in the Simulation Settings window for an AC Sweep, check the Parameter Sweep box to repeat the simulation for each of the three resistor values. Instead of specifying a range of resistor values, create a list of values by clicking the List button in the Parameter Sweep dialog, and entering the list of values, separated by commas or spaces.

After simulation completes, use Probe to create two plots on the screen. In the upper plot, add the trace of the voltage magnitude in

dB at the Out node, which is the magnitude of the capacitor voltage
in dB. The frequency axis is already logarithmic because the AC Sweep
simulation settings specified frequencies in decades. In the lower plot,
add the trace of the voltage phase angle at the Out node, which is the
capacitor's voltage phase angle. Together, the two plots represent the
Bode plot of the capacitor voltage for the three values of resistance.

The RLC circuit's transfer function for this circuit takes the form

$$H(s) = \frac{1/LC}{s^2 + (R/L)s + 1/LC}$$

so has two poles and no zeros. As the value of R changes, the values of
the poles change, as seen here:

R [Ω]	First pole [rad/s]	Second pole [rad/s]
5	$-100 + \text{j}6323.76$	$-100 - \text{j}6323.76$
50	$-1000 + \text{j}6245.00$	$-1000 - \text{j}6245.00$
500	-2254.03	-17745.97

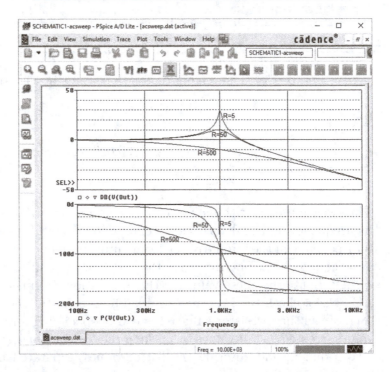

Figure 109: Bode plots for the circuit in Fig. 107, with $R_1 = 5\ \Omega$, $50\ \Omega$, and
$500\ \Omega$.

The two smaller values of R have complex poles with damping coefficients of 0.0158 for $R = 5\ \Omega$ and 0.1581 for $R = 50\ \Omega$. Think about what impact changing resistance has on the shape of the Bode plots for this circuit.

Figure 109 presents the Bode plots for the three resistor values. The magnitude in dB, shown in the upper plot, starts at 0 dB for low frequencies and has a slope of -40 dB/decade for high frequencies. As you might have expected, the dB curve peaks near 1000 Hz for the circuit with the 5 Ω resistor, since it has complex poles and a small damping coefficient.

The lower plot in Figure 109 is the phase angle plot. For each of the three resistor values, the phase angle is $0°$ at low frequencies, $-180°$ at high frequencies, and passes through $-90°$ at the center frequency near 1000 Hz. The steepest phase angle transition occurs for the 5 Ω resistor, because of the complex poles and small damping coefficient.

12.4 FILTER DESIGN

So far we analyzed a circuit's frequency response using PSpice and Probe. We can also employ PSpice and Probe when designing a circuit's frequency response. Example 21 uses PSpice and Probe to verify the frequency domain specifications of a high-Q bandpass filter.

EXAMPLE 21

The circuit in Fig. 110 is an active high-Q bandpass filter. Using 1 nF capacitors and an ideal op amp, design values for the three resistors to yield a center frequency of 10 kHz, a quality factor of 10, and a passband gain of 3. Use PSpice and Probe to verify that the resistor values you calculate produce a filter that meets the three frequency response specifications.

SOLUTION

The resistor design equations for the prototype high-Q filter are

$$R_1 = Q/K = 10/3 = 3.333;$$
$$R_2 = Q/(2Q^2 - K) = 10/197;$$
$$R_3 = 2Q = 20.$$

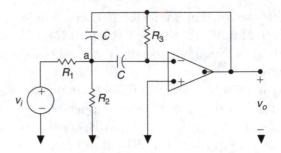

Figure 110: The active high-Q bandpass filter for Example 21.

The scaling factors are $k_f = 2\pi(\omega_o) = 20000\pi$ and $k_m = 1/(Ck_f) = 10^9/k_f$. After scaling,

$$R_1 = 53.1 \text{ k}\Omega;$$
$$R_2 = 808 \text{ }\Omega;$$
$$R_3 = 318.3 \text{ k}\Omega.$$

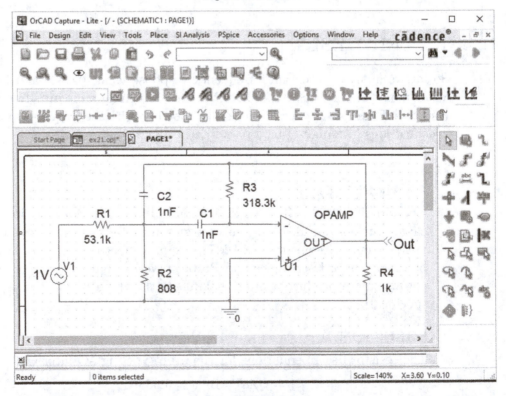

Figure 111: The schematic for the high-Q filter in Fig. 110.

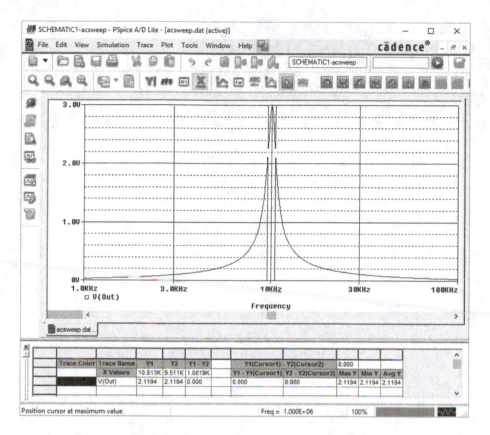

Figure 112: The frequency response for the high-Q filter of Example 21.

The PSpice schematic for the design is shown in Fig. 111. We used the PSpice model for the ideal op amp. The Probe plot of the frequency response after simulation is shown in Fig. 112. Using the cursors, we verify the bandwidth of $10000/10 = 1000$ Hz, and you can see from this plot that the center frequency and passband gain specifications are also satisfied. You can explore the robustness of this design by substituting a realistic op amp model for the ideal model and checking to see if the design specifications are still met.

ASSESSMENT

♦ Use PSpice to find and plot the frequency response for the voltage or current indicated in the following problems, and answer the questions posed: 14.30, 14.31(a)–(d), 15.28(a),(f), and 15.29(a),(f).

Chapter 13

FOURIER SERIES

Both PSpice and Probe can help you understand the frequency spectrum characteristics of periodic waveforms. We begin with a brief discussion of PSpice models for periodic waveforms, then turn to PSpice support for Fourier analysis.

13.1 PULSED SOURCES

The PSpice model for a periodic pulsed voltage source has the part name Vpulse, in the Source library. The pulse parameters are

V1 = initial value (volts);
V2 = pulsed value (volts);
TD = delay time (seconds);
TR = rise time (seconds);
TF = fall time (seconds);
PW = pulse width (seconds);
PER = period (seconds).

Figure 113 plots a periodic pulse and identifies its parameters, which are specified using the Property Editor for the pulsed source.

We analyze a circuit with a pulsed source in the next section.

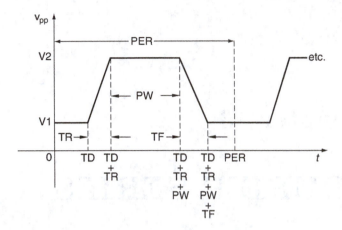

Figure 113: A periodic pulsed voltage source.

13.2 FOURIER ANALYSIS

Transient analysis must precede Fourier series analysis because the Fourier series coefficients are computed for the last complete cycle of the transient analysis waveform. Therefore, the duration of the transient analysis must be at least one cycle of the periodic waveform. If you want PSpice to compute the Fourier series coefficients for the steady-state periodic waveform, transient analysis should be performed for a number of cycles sufficient to ensure the decay of the natural response.

Fourier analysis is specified as an option in transient analysis, as we will see in Example 22. You need to specify the fundamental frequency of the periodic signal in hertz; recall that the fundamental frequency in hertz is the inverse of the signal's period in seconds. You also identify the variables for which the Fourier coefficients are computed. As usual, these variables must be voltages or currents. Finally, you input the number of harmonic terms written to the output file. It is not necessary to include a printer part in your schematic since the Fourier coefficients are automatically printed in the output file during PSpice simulation.

Once PSpice has completed the transient analysis, you can use Probe to calculate and display the Fourier coefficients, whether or not you have specified that the Fourier coefficients be written to the output file. The frequency resolution displayed by Probe is inversely proportional to the transient analysis duration. Therefore, if you want more accurate Fourier coefficients you should specify a longer time for the transient analysis.

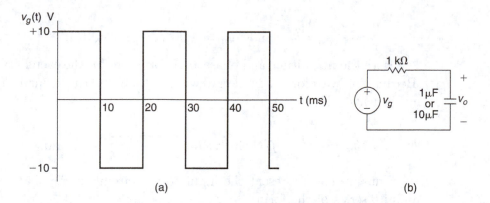

Figure 114: An RC circuit with pulsed voltage input; (a) the pulsed voltage waveform; (b) the circuit.

Example 22 simulates a circuit with a pulsed voltage source, then uses both PSpice and Probe to calculate the Fourier series coefficients.

EXAMPLE 22

Suppose that the pulsed waveform shown in Fig. 114(a) represents the voltage source in the RC circuit shown in Fig. 114(b). Use Fourier series analysis of the input waveform to predict the frequency content of the output waveform for two values of C: 1 μF and 10 μF. Then use the Fourier transform mode in Probe to confirm the PSpice analysis and to display the frequency spectra graphically.

SOLUTION

To begin, calculate the Fourier series coefficients for the pulsed voltage signal shown in Fig. 114(a). Note that this waveform has two properties of symmetry: odd symmetry and half-wave symmetry. These symmetry properties tell us several things about the Fourier coefficients:

- The dc component is 0.

- The Fourier series has only sine terms; the a_n coefficients are 0 for all values of n.

- Only the odd sine terms are nonzero; the b_n coefficients are 0 for even n.

- Because of the form of the a_n's and the b_n's, the Fourier coefficients expressed in polar form are

$$A_n = b_n \quad \text{and} \quad \theta_n = -90°, \quad \text{for odd } n.$$

Both the magnitude and phase angles are zero for the even harmonics. Because of waveform symmetry, we can simplify the formula for the coefficients:

$$b_n = \frac{4}{T} \int_0^{T/2} f(t) \sin k\omega_0 t\, dt = \frac{80}{2k\pi} = \frac{40}{k\pi}, \quad k \text{ odd.}$$

Thus we can represent the input to the circuit in Fig. 114(b) as an infinite series of the form

$$v_g = \frac{40}{\pi} \sin \omega_0 t + \frac{40}{3\pi} \sin 3\omega_0 t + \frac{40}{5\pi} \sin 5\omega_0 t + \ldots$$

The circuit shown in Fig. 114(b) is linear and thus superposition applies. To compute the response of this circuit to the pulsed waveform, we consider the response to each term in the Fourier series independently and sum those responses. From a phasor analysis of the circuit, we know that

$$V_{ok} = \frac{V_{gk}}{1 + jk\omega_0 RC},$$

which we represent in polar form as

$$V_{ok} = \frac{\frac{40}{k\pi} \underline{/0°}}{\sqrt{(k\omega_0 RC)^2 + 1^2} \underline{/\arctan k\omega_0 RC}}.$$

The following table compares the magnitude and phase angle of the input pulsed waveform with the magnitude and phase angle of the voltage across the capacitor for the circuit with $C = 1 \ \mu F$ and the circuit with $C = 10 \ \mu F$, for the first three odd harmonic frequencies. For the circuit with the smaller capacitor, the magnitudes do not change much between the input and the output. For the circuit with the larger capacitor, the magnitudes are much smaller at the output than they are at the input for all harmonic frequencies. We use Probe to investigate this difference further.

Harmonic	V_g	$V_o, C = 1 \ \mu F$	$V_o, C = 10 \ \mu F$
First	$12.7\underline{/0°}$	$12.1\underline{/-17.4°}$	$3.9\underline{/-72.3°}$
Third	$4.2\underline{/0°}$	$3.1\underline{/-43.3°}$	$0.4\underline{/-83.9°}$
Fifth	$2.5\underline{/0°}$	$1.4\underline{/-57.5°}$	$0.2\underline{/-86.4°}$

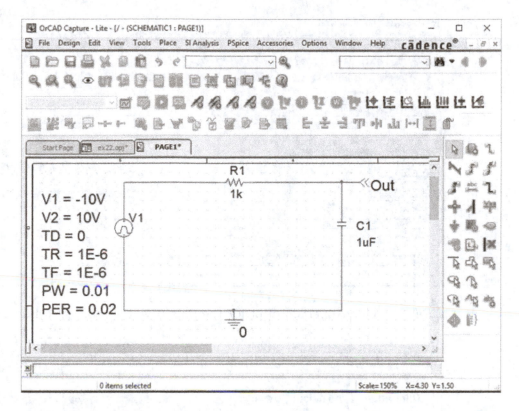

Figure 115: The schematic for the circuit in Fig. 114(b), with $C = 1 \ \mu$F.

The PSpice schematic for the circuit in Fig. 114(b) with the smaller capacitor value and the input shown in Fig. 114(a) is shown in Fig. 115. The Property Editor used to specify the pulsed voltage source parameters is shown in Fig. 116. You should confirm that the values of the properties match the pulsed waveform in Fig. 114(a).

We specify the parameters for transient analysis using the Simulation Settings. The value for TSTOP is based on the time constant for this RC circuit, which for $C = 1 \ \mu$F is 1 ms and for $C = 10 \ \mu$F is 10 ms. Thus, setting TSTOP to 60 ms will allow a 6 time-constant decay of the natural response for the larger value of C, ensuring that the Fourier series coefficients will be computed for the steady-state waveform.

We direct PSpice to compute the Fourier coefficients and write them to the output file by clicking the Output File Options button in the Transient Analysis Simulation Settings window. Then check the box to perform the Fourier analysis and specify the fundamental frequency, the number of harmonics, and the circuit variables. The completed dialog for Fourier analysis is shown in Fig. 117.

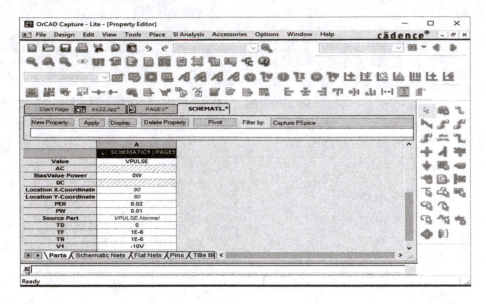

Figure 116: The Property Editor for the pulsed voltage source in Fig. 115.

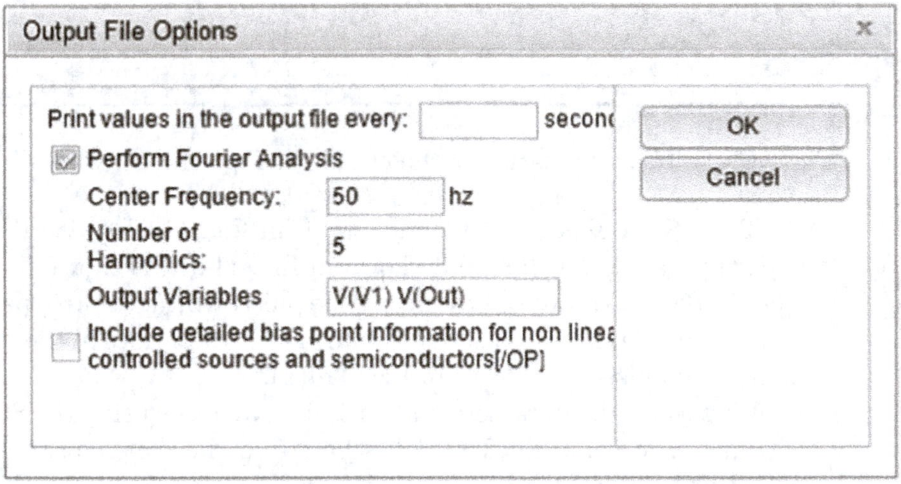

Figure 117: The Fourier analysis dialog in Transient analysis.

The Fourier coefficients for the first five harmonics for the output capacitor voltage are shown in Fig. 118. Note that this PSpice analysis confirms our previous hand calculations for the first three odd harmonics. Note further that the even harmonics are essentially zero; the small nonzero values that are printed arise because of the imprecision inherent in digital computer calculations.

Now examine the graph of the time response for both the input

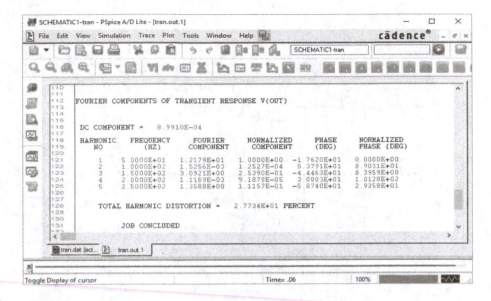

Figure 118: The first five harmonics in the output voltage for the schematic in Fig. 115.

pulsed waveform and the output capacitor voltage produced by Probe and shown in Fig. 119. Because of the small capacitor value, the output voltage follows the input voltage. Finding that the Fourier coefficients of the input and output voltages are so similar is not surprising.

Figure 120, produced from Fig. 119 in Probe by clicking the Toolbar button labeled FFT and changing the range of the x-axis, confirms this similarity in a single plot.

We changed the capacitor value in the schematic (Fig. 115) to 10 μF, and repeated the analysis. The resulting Fourier series coefficients for the first five harmonics of the capacitor voltage are shown in Fig. 121, confirming our earlier calculations.

Probe plots of the time response for the input and output voltages, shown in Fig. 122, show that the large capacitor value prevents the output voltage from following the input voltage. Because the voltage waveforms are so different in the time domain, you should expect very different Fourier coefficients. You saw this difference in both the hand calculations and the PSpice analysis. The Probe-generated plot of the Fourier transformed voltages shown in Fig. 123 further confirms the different spectral content in the two waveforms.

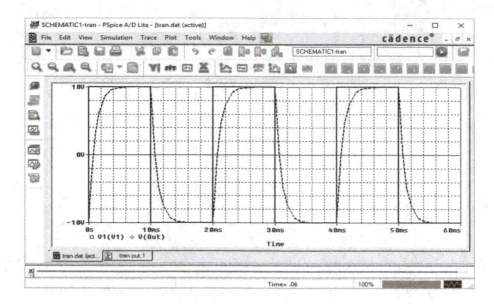

Figure 119: Comparing the input and out voltage for the schematic in Fig. 115.

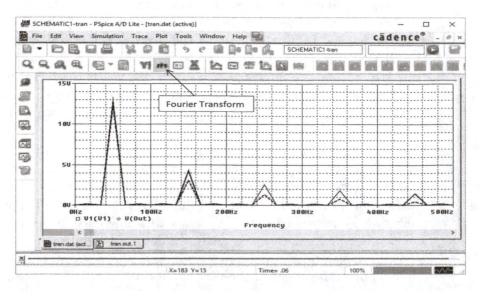

Figure 120: Fourier transform of the input and output voltages in Fig. 119.

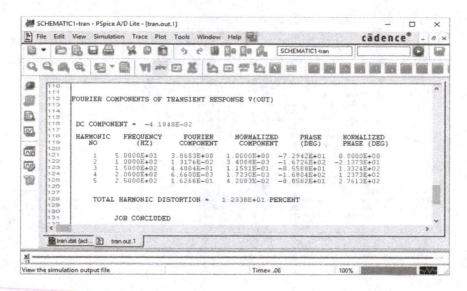

Figure 121: The first five harmonics in the output voltage for the RC circuit in Fig. 114(b) with $C = 10 \ \mu\text{F}$.

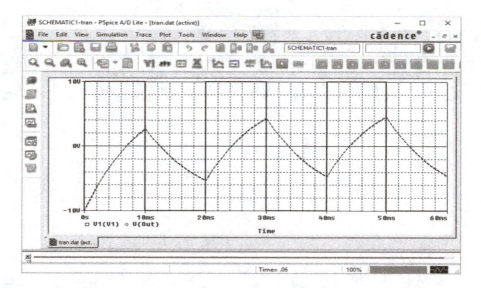

Figure 122: Comparing the input and out voltage for the RC circuit in Fig. 114(b) with $C = 10 \ \mu\text{F}$.

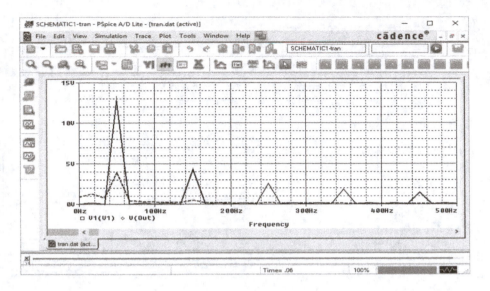

Figure 123: Fourier transform of the input and output voltages in Fig. 122.

ASSESSMENT

♦ Use PSpice to find and plot the Fourier coefficients for the voltages and currents indicated in the following problems: 16.28 and 16.29.

Chapter 14

SUMMARY

OrCAD PSpice is designed to simulate an electric circuit or network. The user defines a circuit in terms of a topological, as opposed to a mathematical, description. Although the primary purpose of PSpice is to simulate circuits containing nonlinear electronic components, we have introduced it to you as a tool for analyzing linear, lumped-parameter circuits. There are several reasons for this early introduction. First, it acquaints you with a powerful computational tool. Second, you can test the computer-generated values using circuit analysis techniques, thus gaining confidence in your circuit analysis skills. Third, you can simulate circuits that are complex and would be difficult to analyze without the computer's support. Such complex problems can add significantly to your understanding of circuit behavior. Finally, simulation software is part of modern engineering practice. Introducing this software at an elementary level is a first step in preparing you for your engineering career.

TO EXPLORE FURTHER

Additional information is available by using the Help menus available in ORCAD Lite PSpice and by exploring the extensive tutorials available here:

http://www.orcad.com/resources/orcad-tutorials.

APPENDIX

QUICK REFERENCE TO PSPICE NETLIST STATEMENTS

Before PSpice can begin analysis, the schematic you create is translated into a collection of statements that identify the components, their attributes, and their interconnections. Collectively, these statements define the schematic's netlist. This appendix presents the netlist syntax for most of the schematic parts presented in this supplement. Familiarity with the netlist syntax can help you debug errors in your schematic.

DATA STATEMENTS

In the syntax for data statements, *xxx* represents any alphanumeric string that uniquely identifies a circuit component, n+ represents the positive component node, and n− represents the negative component node. Positive voltage drop is from n+ to n−, and positive current flows from n+ to n−. Optional elements in the syntax specification are enclosed by brackets ([]).

INDEPENDENT DC VOLTAGE AND CURRENT SOURCES

SYNTAX (VOLTAGE)	Vxxx n+	n−	DC	*value*
SYNTAX (CURRENT)	Ixxx n+	n−	DC	*value*

where *value* is the dc voltage in volts for the voltage source and the dc current in amperes for the current source.

DESCRIPTION Provides a constant source of voltage or current to the circuit.

EXAMPLES   V_in $N_0001 $N_0002 DC 3.5

I_source $N_0003 0 DC 0.25

INDEPENDENT AC VOLTAGE AND CURRENT SOURCES

SYNTAX (VOLTAGE)   Vxxx   n+   n−   AC   *mag*   [*phase*]

SYNTAX (CURRENT)   Ixxx   n+   n−   AC   *mag*   [*phase*]

where

mag is the magnitude of the ac waveform in volts for the voltage source and in amperes for the current source;

phase is the phase angle of the ac waveform in degrees and has a default value of 0.

DESCRIPTION Supplies a sinusoidal voltage or current at a fixed frequency to the circuit.

EXAMPLES   V_ab $N_0003 $N_0002 AC 10 90

I_12 $N_0004 $N_0006 AC 0.5

INDEPENDENT TRANSIENT VOLTAGE AND CURRENT SOURCES

SYNTAX (VOLTAGE)   Vxxx   n+   n−   *trans_type*

SYNTAX (CURRENT)   Ixxx   n+   n−   *trans_type*

where *trans_type* is one of the following transient waveform types:

- **EXP** (*start peak* [*delay1*] [*rise*] [*delay2*] [*fall*])

  This is an exponential waveform where

  start is the initial value of voltage in volts or current in amperes;

  peak is the maximum value of voltage in volts or current in amperes;

  delay1 is the delay in seconds prior to a change in voltage or current from the initial value, which is 0 seconds by default;

rise is the time constant in seconds of the exponential decay from start to peak, which has a default value of *tstep* seconds (see .TRAN);

delay2 is the delay in seconds prior to introducing a second exponential decay, from the maximum value toward the initial value, which has a default value of *delay1* + *tstep* seconds (see .TRAN);

fall is the time constant in seconds of the exponential decay from *peak* to final, which has a default value of *tstep* seconds (see .TRAN).

- PULSE(*min max* [*delay*] [*rise*] [*fall*] [*width*] [*period*])

  This is a pulse waveform where

  min is the minimum value of the waveform in volts for the voltage source and in amperes for the current source;

  max is the maximum value of the waveform in volts for the voltage source and in amperes for the current source;

  delay is the time in seconds prior to the onset of the pulse train, which has a default of 0 seconds;

  rise is the time in seconds for the waveform to transition from min to max, which has a default value of *tstep* seconds (see .TRAN);

  fall is the time in seconds for the waveform to transition from *max* to *min*, which has a default value of *tstep* seconds (see .TRAN);

  width is the time in seconds that the waveform remains at the max value, which has a default value of *tstop* seconds (see .TRAN);

  period is the time in seconds that separates the pulses in the pulse train, which has a default of *tstop* seconds (see .TRAN).

- PWL(*t1 val1 t2 val2* . . .)

  This is a piecewise linear waveform where

  t1 val1 are a paired time (in seconds) and value (in volts for a voltage source and amperes for a current source) that specifies a corner of the waveform; all pairs of times and values are linearly connected to form the whole waveform.

- SIN(*off peak* [*freq*] [*delay*] [*damp*] [*phase*])

This is a damped sinusoidal waveform where

off is the initial value of the voltage in volts or the current in amperes;

peak is the maximum amplitude of the voltage in volts or the current in amperes;

freq is the sinusoidal frequency in hertz, which has a default value of $1/tstop$ hertz (see .TRAN);

delay is the time in seconds that the waveform stays at the initial value before the sinusoidal oscillation begins, which has a default value of 0 seconds;

damp is the sinusoidal damping factor in seconds^{-1}used to specify the decaying exponential envelope for the sinusoid, which has a default value of 0 seconds^{-1};

phase is the initial phase angle of the sinusoidal waveform in degrees, which has a default value of 0 degrees.

DESCRIPTION Supplies a time-varying voltage or current to the circuit whose waveform can be characterized as exponential, pulsed, piecewise linear, or damped sinusoidal.

EXAMPLES

V_s $N_0003 $N_0001 EXP(2 6 .5 .1 .5 .2)

I_in $N_0004 0 PULSE(-.3 .3 0 .01 .01 1 2)

I_5 $N_0002 $N_0003 PWL(0 .2 1 .6 1.5 6 3 -.5)

V_g $N_0003 SIN(0 2 100 0 0 90)

DEPENDENT VOLTAGE-CONTROLLED VOLTAGE SOURCE

SYNTAX Exxx   n+   n−   cn+   cn−   *gain*

where

cn+ is the positive node for the controlling voltage;

cn− is the negative node for the controlling voltage;

gain is the ratio of the source voltage (between n+ and n −) to the controlling voltage (between *cn+* and *cn−*).

DESCRIPTION Provides a voltage source whose value depends on a voltage measured elsewhere in the circuit.

EXAMPLE E_op $N_0001 $N_0002 $N_0004   0   .5

DEPENDENT CURRENT-CONTROLLED CURRENT SOURCE

SYNTAX Fxxx   n+   n−   *Vyyy*   *gain*
>
> where
>
>> *Vyyy* is the name of the voltage source through which the controlling current flows;
>>
>> *gain* is the ratio of the source current (flowing from n+ to n−) to the controlling current (flowing through *Vyyy*).

DESCRIPTION Provides a current source whose value depends on the magnitude of a current flowing through a voltage source elsewhere in the circuit.

EXAMPLE F_dep $N_0003 $N_0002   V_control   10

DEPENDENT VOLTAGE-CONTROLLED CURRENT SOURCE

SYNTAX Gxxx   n+   n−   *cn*+   *cn*−   *gain*
>
> where
>
>> *cn*+ is the positive node for the controlling voltage;
>>
>> *cn*− is the negative node for the controlling voltage;
>>
>> *gain* is the ratio of the source current (flowing from n+ to n−) to the controlling voltage (between *cn*+ and *cn*−), in ohms^{-1}.

DESCRIPTION Provides a current source whose value depends on the magnitude of a voltage measured elsewhere in the circuit.

EXAMPLE G_on $N_0003 $N_0006 $N_0004 $N_0001   0.35

DEPENDENT CURRENT-CONTROLLED VOLTAGE SOURCE

SYNTAX Hxxx   n+   n−   *Vyyy*   *gain*
>
> where
>
>> *Vyyy* is the name of the voltage source through which the controlling current flows;
>>
>> *gain* is the ratio of the source voltage (between n+ and n−) to the controlling current (flowing through *Vyyy*), in ohms.

DESCRIPTION Provides a voltage source whose value depends on the magnitude of current flowing through another voltage source elsewhere in the circuit.

EXAMPLE H_out $N_0007 $N_0002   V_dummy   −2.5E−3

RESISTOR
SYNTAX Rxxx   n+   n−   [*mname*]   *value*

> where

> > *mname* is the name of a RES model defined in a .MODEL statement (see .MODEL)—note that *mname* is optional;

> > *value* is the resistance, in ohms.

DESCRIPTION Models a resistor, a circuit element whose voltage and current are linearly dependent.

EXAMPLE R_for $N_0003 $N_0004   16E3

INDUCTOR
SYNTAX Lxxx   n+   n−   [*mname*]   *value*   [IC = *icval*]

> where

> > *mname* is the name of an IND model defined in a .MODEL statement (see .MODEL)—note that *mname* is optional;

> > *value* is the inductance, in henries;

> > *icval* is the initial value of the current in the inductor, in amperes—note that specifying the initial condition is optional.

DESCRIPTION Models an inductor, a circuit element whose voltage is linearly dependent on the derivative of its current.

EXAMPLE L_44 $N_0001 $N_0009   3E−3 IC = 1E−2

CAPACITOR
SYNTAX Cxxx   n+   n−   [*mname*]   *value*   [IC = *icval*]
> where
> > *mname* is the name of a CAP model defined in a .MODEL statement (see .MODEL)—note that *mname* is optional;

value is the capacitance, in farads;

icval is the initial value of the voltage across the capacitor, in volts—note that specifying the initial condition is optional.

DESCRIPTION Models a capacitor, a circuit element whose current is linearly dependent on the derivative of its voltage.

EXAMPLES   C_two $N_0004   0   cmod   2E−6

.MODEL   cmod   CAP(C = 1)

MUTUAL INDUCTANCE

SYNTAX Kxxx   *Lyyy*   *Lzzz*   *value*

where

Lyyy is the name of the inductor on the primary side of the coil;

Lzzz is the name of the inductor on the secondary side of the coil;

value is the mutual coupling coefficient, k, which has a value such that $0 \leq k < 1$.

DESCRIPTION Models the magnetic coupling between any two inductor coils in a circuit.

EXAMPLE K_ab   L_a   L_b   0.5

SUBCIRCUIT DEFINITION

SYNTAX .SUBCKT   *name*   [*nodes*]   .ENDS

where

name is the name of the subcircuit, as referenced by an X statement (see **subcircuit call**);

nodes is the optional list of nodes used to identify the connections to the subcircuit;

.ENDS signifies the end of the subcircuit definition.

DESCRIPTION Used to provide "subroutine"-type definitions of portions of a circuit. When the subcircuit is referenced in an X statement, the definition between the .SUBCKT statement and the .ENDS statement replaces the X statement in the source file.

EXAMPLES

```
.SUBCKT  opamp  $N_0001  $N_0002  $N_0003  $N_0004  $N_0005
.ENDS
```

SUBCIRCUIT CALL

SYNTAX Xxxx [*nodes*] *name*

> where
>> *nodes* is the optional list of circuit nodes used to connect the subcircuit into the rest of the circuit; there must be as many nodes in this list as there are in the subcircuit definition (see "Subcircuit Definition");
>> *name* is the name of the subcircuit as defined in a .SUBCKT statement.

DESCRIPTION Replaces the X statement with the definition of a subcircuit, which permits a subcircuit to be defined once and used many times within a given source file.

EXAMPLE

```
X_amp  $N_0004  $N_0002  $N_0008  $N_0005  $N_0001  opamp
```

LIBRARY FILE

SYNTAX .LIB [*fname*]

> where *fname* is the name of the library file containing .MODEL or .SUBCKT statements referenced in the source file, which by default is the nominal or evaluation library file.

DESCRIPTION Used to reference models or subcircuits.

EXAMPLE .LIB mylib.lib

DEVICE MODELS

SYNTAX .MODEL *mname* *mtype* [(*par=value*)]

> where
>> *mname* is a unique model name, which is also used in the device statement that incorporates this model;

mtype is one of the model types available;

par=value is an optionally specified list of parameters and their assigned values, specific to the model type:

- RES, which models a resistor and has the parameter R, the resistance multiplier, whose default value is 1;

- IND, which models an inductor and has the parameter L, the inductance multiplier, whose default value is 1;

- CAP, which models a capacitor and has the parameter C, the capacitance multiplier, whose default value is 1.

DESCRIPTION Defines standard devices that can be used in a circuit and sets parameter values that characterize the specific device being modeled.

EXAMPLES   .MODEL lmodel IND(L = 2)

               .MODEL resist RES

CONTROL STATEMENTS

DC ANALYSIS

SYNTAX .DC *[type]* *vname* *start* *end* *incre* *[nest-sweep]*

where

type is the sweep type and must be LIN for a linear sweep from the *start* value to the *end* value, OCT for a logarithmic sweep from *start* value to *end* value in octaves, DEC for a logarithmic sweep from *start* value to *end* value in decades, or LIST if a list of values is to be used; the default sweep type is LIN;

vname is the name of the circuit element whose value is swept, usually an independent voltage or current source;

start is the beginning value of *vname*;

end is the final value of *vname*; note that when the sweep type is LIST, the *start*, *end*, and *incre* values are replaced by a list of values that *vname* will take on;

incre is the step size for the LIN sweep type and is the number of points per octave or decade for the OCT and DEC sweep types;

nest-sweep is an optional additional sweep of a second circuit variable, which follows the same syntax as the sweep for the first variable.

DESCRIPTION Provides a method for varying one or two dc source values and analyzing the circuit for each sweep value.

EXAMPLES `.DC  V_ab  $N_0001  $N_0010  .5`

`.DC  I_1  DEC  .1  10  30  V_1  DEC  10  100  30`

SENSITIVITY

SYNTAX `.SENS` *vname*

where *vname* is the circuit variable name (or list of circuit variable names) for which sensitivity analysis will be performed.

DESCRIPTION Computes and prints to the output file a dc sensitivity analysis of vname to the values of other circuit elements, including resistors, independent voltage and current sources, and voltage- and current-controlled switches.

EXAMPLE `.SENS  V($N_0002)  I(V_in)`

TRANSIENT ANALYSIS

SYNTAX `.TRAN` *tstep* *tstop* [*npval*] *istep* [UIC]

where

tstep is the time interval separating values that are printed or plotted;

tstop is the ending time for the transient analysis;

npval is the time between 0 and the first value printed or plotted, which by default is 0;

istep is the internal time step used for computing values, which by default is *tstop*/50;

UIC will bypass the calculation of the bias point, which usually precedes the transient analysis, and use the initial conditions specified by IC = in inductor and capacitor data statements instead.

DESCRIPTION Performs a transient analysis of the circuit described in the source file to calculate the values of circuit variables as a function of time.

EXAMPLE .TRAN .01 10 0 .001 UIC

INITIAL CONDITION

SYNTAX .IC *Vnode=value*

> where
>
> *Vnode=value* is one or a list of pairs consisting of a voltage node *Vnode*, represented in standard form, and the initial value in volts at that node.

DESCRIPTION Sets the initial conditions for transient analysis by specifying one or more node voltage values.

EXAMPLE .IC V($N_0001) = 10

AC ANALYSIS

SYNTAX .AC [*type*] *num start end*

> where
>
> *type* is the type of sweep, which must be one of the following keywords:
>
>> LIN for a linear sweep in frequency, which is the default;
>>
>> OCT for a logarithmic sweep in frequency by octaves;
>>
>> DEC for a logarithmic sweep in frequency by decades;
>
> *num* is the total number of points in the sweep for a linear sweep and the number of points per octave or decade for a logarithmic sweep;
>
> *start* is the starting frequency, in hertz;
>
> *end* is the final frequency, in hertz.

DESCRIPTION Computes the frequency response of the circuit described in the source file as the frequency is swept either linearly or logarithmically from an initial value to a final value.

EXAMPLE .AC LIN 300 10 1000

STEPPED VALUES

SYNTAX .STEP [*type*] *name start end incre*

> where
>
> *type* is one of the following keywords describing the type of parameter variation:

LIN for a linear variation, which is the default;

OCT for a logarithmic variation by octaves;

DEC for a logarithmic variation by decades;

name is the name of the circuit element whose value is to be varied;

start is the starting value for the circuit element, in units appropriate to the type of circuit element;

end is the final value for the circuit element, in units appropriate to the type of circuit element;

incre is the step size for a linear variation and the number of values per octave or decade for the logarithmic variation. Note that when a discrete list of parameter values is desired, *type* is not used, and the name is followed by the keyword LIST and a list of parameter values.

DESCRIPTION Used to step a circuit element's value through a range either linearly or logarithmically or through a discrete list, analyzing the circuit for each value.

EXAMPLES   .STEP DEC RES RMOD(R) 10 10E4 3

                .STEP V_2 LIST 1 4 12 19

FOURIER SERIES ANALYSIS

SYNTAX .FOUR *freq   vname*

where

freq is the fundamental frequency of the circuit;

vname is the variable name or the list of variable names for which Fourier series coefficients will be computed.

DESCRIPTION Uses the results of a transient analysis to compute the Fourier coefficients for the first nine harmonics. Note the .FOUR statement requires a .TRAN statement to perform the transient analysis.

EXAMPLE .FOUR 10E3 V($N_0003)

OUTPUT STATEMENTS

OPERATING POINT
SYNTAX .OP

DESCRIPTION Outputs information describing the bias point computations for the circuit being simulated.

EXAMPLE .OP

PRINT RESULTS
SYNTAX .PRINT *type* *vname*

> where
>
> *type* identifies the type of analysis performed to generate the data being printed and must be one of the keywords DC, AC, or TRAN;
>
> *vname* is the variable name or the list of variable names for which values are to be printed.

DESCRIPTION Prints the results of circuit analysis in table form to an output file for each program variable specified.

EXAMPLE .PRINT AC V(R_in) I(V_source)

PROBE

SYNTAX .PROBE [*vname*]

> where *vname* is the optionally specified variable or list of variables whose values from dc, ac, and transient analysis will be stored in the file that PROBE uses to generate plots. If no variable name is included, values of all circuit variables will be stored in the PROBE file.

DESCRIPTION Generates a file of data from dc, ac, and transient analysis used by PROBE to generate high-quality plots.

EXAMPLE .PROBE V($N_0002) I(R_in) I(R_out)

MISCELLANEOUS STATEMENTS

TITLE LINE

DESCRIPTION Each source file must have a title line as its first
line.

EXAMPLE A Circuit to Simulate RLC Frequency Response

END STATEMENT
SYNTAX .END

DESCRIPTION Each source file must have an .END statement as its final
line.

EXAMPLE .END

INDEX